Flinn Scientific
ChemTopic™ Labs

Electrochemistry

Senior Editor

Irene Cesa
Flinn Scientific, Inc.
Batavia, IL

Curriculum Advisory Board

Bob Becker
Kirkwood High School
Kirkwood, MO

Kathleen J. Dombrink
McCluer North High School
Florissant, MO

Robert Lewis
Downers Grove North High School
Downers Grove, IL

John G. Little
St. Mary's High School
Stockton, CA

Lee Marek
University of Illinois–Chicago
Chicago, IL

John Mauch
Braintree High School
Braintree, MA

Dave Tanis
Grand Valley State University
Allendale, MI

FLINN SCIENTIFIC INC.
"Your Safer Source for Science Supplies"
P.O. Box 219 • Batavia, IL 60510
1-800-452-1261 • www.flinnsci.com

ISBN 1-877991-85-6

Copyright © 2005 Flinn Scientific, Inc.

All rights reserved. No part of this book may be reproduced or transmitted in any form or by any means, electronic or mechanical, including, but not limited to photocopy, recording, or any information storage and retrieval system, without permission in writing from Flinn Scientific, Inc. No part of this book may be included on any Web site.

Reproduction permission is granted only to the science teacher who has purchased this volume of Flinn ChemTopic™ Labs, Electrochemistry, Catalog No. AP6662 from Flinn Scientific, Inc. Science teachers may make copies of the reproducible student pages for use only by their students.

Printed in the United States of America.

Table of Contents

	Page
Flinn ChemTopic™ Labs Series Preface	i
About the Curriculum Advisory Board	ii
Electrochemistry Preface	iii
Format and Features	iv–v
Experiment Summaries and Concepts	vi–vii

Experiments

Introduction to Electrochemistry	1
Measuring Cell Potentials	11
Quantitative Electrochemistry	23
Electrolysis Reactions	35

Demonstrations

Microscale Electrolysis	49
Hoffman Electrolysis	53
The Tin Man	57
Orange Juice Clock	59
Basic Electrophoresis	61
Lemon Battery Contest	63

Supplementary Information

Safety and Disposal Guidelines	66
National Science Education Standards	68
Master Materials Guide	70

Flinn ChemTopic™ Labs Series Preface

Lab Manuals Organized Around Key Content Areas in Chemistry

In conversations with chemistry teachers across the country, we have heard a common concern. Teachers are frustrated with their current lab manuals, with experiments that are poorly designed and don't teach core concepts, with procedures that are rigid and inflexible and don't work. Teachers want greater flexibility in their choice of lab activities. As we further listened to experienced master teachers who regularly lead workshops and training seminars, another theme emerged. Master teachers mostly rely on collections of experiments and demonstrations they have put together themselves over the years. Some activities have been passed on like cherished family recipe cards from one teacher to another. Others have been adapted from one format to another to take advantage of new trends in microscale equipment and procedures, technology innovations, and discovery-based learning theory. In all cases the experiments and demonstrations have been fine-tuned based on real classroom experience.

Flinn Scientific has developed a series of lab manuals based on these "cherished recipe cards" of master teachers with proven excellence in both teaching students and training teachers. Created under the direction of an Advisory Board of award-winning chemistry teachers, each lab manual in the Flinn ChemTopic™ Labs series contains 4–6 student-tested experiments that focus on essential concepts and applications in a single content area. Each lab manual also contains 4–6 demonstrations that can be used to illustrate a chemical property, reaction, or relationship and will capture your students' attention. The experiments and demonstrations in the Flinn ChemTopic™ Labs series are enjoyable, highly focused, and will give students a real sense of accomplishment.

Laboratory experiments allow students to experience chemistry by doing chemistry. Experiments have been selected to provide students with a crystal-clear understanding of chemistry concepts and encourage students to think about these concepts critically and analytically. Well-written procedures are guaranteed to work. Reproducible data tables teach students how to organize their data so it is easily analyzed. Comprehensive teacher notes include a master materials list, solution preparation guide, complete sample data, and answers to all questions. Detailed lab hints and teaching tips show you how to conduct the experiment in your lab setting and how to identify student errors and misconceptions before students are led astray.

Chemical demonstrations provide another teaching tool for seeing chemistry in action. Because they are both visual and interactive, demonstrations allow teachers to take students on a journey of observation and understanding. Demonstrations provide additional resources to develop central themes and to magnify the power of observation in the classroom. Demonstrations using discrepant events challenge student misconceptions that must be broken down before new concepts can be learned. Use demonstrations to introduce new ideas, illustrate abstract concepts that cannot be covered in lab experiments, and provide a spark of excitement that will capture student interest and attention.

Safety, flexibility, and choice

Safety always comes first. Depend on Flinn Scientific to give you upfront advice and guidance on all safety and disposal issues. Each activity begins with a description of the hazards involved and the necessary safety precautions to avoid exposure to these hazards. Additional safety, handling, and disposal information is also contained in the teacher notes.

The selection of experiments and demonstrations in each Flinn ChemTopic™ Labs manual gives you the flexibility to choose activities that match the concepts your students need to learn. No single teacher will do all of the experiments and demonstrations with a single class. Some experiments and demonstrations may be more helpful with a beginning-level class, while others may be more suitable with an honors class. All of the experiments and demonstrations have been keyed to national content standards in science education.

Chemistry is an experimental science!

Whether they are practicing key measurement skills or searching for trends in the chemical properties of substances, all students will benefit from the opportunity to discover chemistry by doing chemistry. No matter what chemistry textbook you use in the classroom, Flinn ChemTopic™ Labs will help you give your students the necessary knowledge, skills, attitudes, and values to be successful in chemistry.

About the Curriculum Advisory Board

Flinn Scientific is honored to work with an outstanding group of dedicated chemistry teachers. The members of the Flinn ChemTopic Labs Advisory Board have generously contributed their proven experiments and demonstrations to create these topic lab manuals. The wisdom, experience, creativity, and insight reflected in their lab activities guarantee that students who perform them will be more successful in learning chemistry. On behalf of all chemistry teachers, we thank the Advisory Board members for their service and dedication to chemistry education.

Bob Becker teaches chemistry and AP chemistry at Kirkwood High School in Kirkwood, MO. Bob received his B.A. from Yale University and M.Ed. from Washington University and has 20 years of teaching experience. A well-known demonstrator, Bob has conducted more than 100 demonstration workshops across the U.S. and Canada and was a Team Leader for the Flinn Foundation Summer Workshop Program. His creative and unusual demonstrations have been published in the *Journal of Chemical Education,* the *Science Teacher,* and *Chem13 News.* Bob is the author of two books of chemical demonstrations, *Twenty Demonstrations Guaranteed to Knock Your Socks Off, Volumes I and II,* published by Flinn Scientific. Bob has been awarded the James Bryant Conant Award in High School Teaching from the American Chemical Society, the Regional Catalyst Award from the Chemical Manufacturers Association, and the Tandy Technology Scholar Award.

Kathleen J. Dombrink teaches chemistry and advanced-credit college chemistry at McCluer North High School in Florissant, MO. Kathleen received her B.A. in Chemistry from Holy Names College and M.S. in Chemistry from St. Louis University and has 35 years of teaching experience. Recognized for her strong support of professional development, Kathleen has been selected to participate in the Fulbright Memorial Fund Teacher Program in Japan and NEWMAST and Dow/NSTA Workshops. She served as co-editor of the inaugural issues of *Chem Matters* and was a Woodrow Wilson National Fellowship Foundation Chemistry Team Member for 11 years. Kathleen is currently a Team Leader for the Flinn Foundation Summer Workshop Program. Kathleen has received the Presidential Award, the Midwest Regional Teaching Award from the American Chemical Society, the Tandy Technology Scholar Award, and a Regional Catalyst Award from the Chemical Manufacturers Association.

Robert Lewis retired from teaching chemistry at Downers Grove North High School in Downers Grove, IL, and is currently a Secondary Coordinator for the GATE program in Chicago. Robert received his B.A. from North Central College and M.A. from University of the South and has 30 years of teaching experience. He was a founding member of Weird Science, a group of chemistry teachers that traveled throughout the country to stimulate teacher enthusiasm for using demonstrations to teach science. Robert served as a Team Leader for both the Woodrow Wilson National Fellowship Foundation and the Flinn Foundation Summer Workshop Program. Robert has received the Presidential Award, the James Bryant Conant Award in High School Teaching from the American Chemical Society, the Tandy Technology Scholar Award, a Regional Catalyst Award from the Chemical Manufacturers Association, and a Golden Apple Award from the State of Illinois.

John G. Little teaches chemistry and AP chemistry at St. Mary's High School in Stockton, CA. John received his B.S. and M.S. in Chemistry from University of the Pacific and has 39 years of teaching experience. Highly respected for his well-designed labs, John is the author of two lab manuals, *Chemistry Microscale Laboratory Manual* (D. C. Heath), and *Microscale Experiments for General Chemistry* (with Kenneth Williamson, Houghton Mifflin). He is also a contributing author to *Science Explorer* (Prentice Hall) and *World of Chemistry* (McDougal Littell). John served as a Chemistry Team Leader for both the Woodrow Wilson National Fellowship Foundation and the Flinn Foundation Summer Workshop Program. He has been recognized for his dedicated teaching with the Tandy Technology Scholar Award and the Regional Catalyst Award from the Chemical Manufacturers Association.

Lee Marek retired from teaching chemistry at Naperville North High School in Naperville, IL and currently teaches at the University of Illinois–Chicago. Lee received his B.S. in Chemical Engineering from the University of Illinois and M.S. degrees in Physics and Chemistry from Roosevelt University. He has more than 30 years of teaching experience and is a Team Leader for the Flinn Foundation Summer Workshop Program. His students have won national recognition in the International Chemistry Olympiad, the Westinghouse Science Talent Search, and the Internet Science and Technology Fair. Lee was a founding member of Weird Science and has presented more than 500 demonstration and teaching workshops for more than 300,000 students and teachers across the country. Lee has performed science demonstrations on the *David Letterman Show* 20 times. Lee has received the Presidential Award, the James Bryant Conant Award in High School Teaching and the Helen M. Free Award for Public Outreach from the American Chemical Society, the National Catalyst Award from the Chemical Manufacturers Association, and the Tandy Technology Scholar Award.

John Mauch teaches chemistry and AP chemistry at Braintree High School in Braintree, MA. John received his B.A. in Chemistry from Whitworth College and M.A. in Curriculum and Education from Washington State University and has more than 25 years of teaching experience. John is an expert in microscale chemistry and is the author of two lab manuals, *Chemistry in Microscale, Volumes I and II* (Kendall/Hunt). He is also a dynamic and prolific demonstrator and workshop leader. John has presented the Flinn Scientific Chem Demo Extravaganza show at NSTA conventions for eight years and has conducted more than 100 workshops across the country. John was a Chemistry Team Member for the Woodrow Wilson National Fellowship Foundation program and is currently a Board Member for the Flinn Foundation Summer Workshop Program. John has received the Massachusetts Chemistry Teacher of the Year Award from the New England Institute of Chemists.

Dave Tanis is Associate Professor of Chemistry at Grand Valley State University in Allendale, MI. Dave received his B.S. in Physics and Mathematics from Calvin College and M.S. in Chemistry from Case Western Reserve University. He taught high school chemistry for 26 years before joining the staff at Grand Valley State University to direct a coalition for improving pre-college math and science education. Dave later joined the faculty at Grand Valley State University and currently teaches courses for pre-service teachers. The author of two laboratory manuals, Dave acknowledges the influence of early encounters with Hubert Alyea, Marge Gardner, Henry Heikkinen, and Bassam Shakhashiri in stimulating his long-standing interest in chemical demonstrations and experiments. Continuing this tradition of mentorship, Dave has led more than 40 one-week institutes for chemistry teachers and served as a Team Member for the Woodrow Wilson National Fellowship Foundation for 13 years. He is currently a Board Member for the Flinn Foundation Summer Workshop Program. Dave received the College Science Teacher of the Year Award from the Michigan Science Teachers Association.

Preface
Electrochemistry

Principles of electricity and chemistry overlap in electrochemistry, the study of the interconversion of chemical and electrical energy in chemical reactions. What are the basic features of an electrochemical cell? What chemical reactions will take place when an external voltage is applied to an electrolytic cell? What factors determine the ability of a voltaic cell to produce electricity? What is the relationship between the amount of electricity that flows through a solution and the extent of the chemical reaction that takes place? The purpose of *Electrochemistry,* Volume 17 in the Flinn ChemTopic™ Labs series, is to provide high school chemistry teachers with laboratory activities that will help students understand the basic principles of electrochemistry and its applications. Four experiments and six demonstrations allow students to build simple electrochemical cells and to investigate the production and utilization of electricity. Please see *Oxidation and Reduction,* Volume 16 in the Flinn ChemTopic™ Labs series, for activities dealing with the fundamental role of electron transfer in oxidation and reduction reactions.

Basic Principles

The basic features of an electrochemical cell are explored for both electrolytic and voltaic cells. In the experiment "Introduction to Electrochemistry," students study the electrolysis of water in a simple U-tube using mechanical pencil "lead" electrodes and a 9-V battery. Indicator color changes reveal the oxidation and reduction reactions that take place at each electrode and the properties of the anode and the cathode. In "Measuring Cell Potentials," students build unique micro-voltaic cells using metals and metal ion solutions and measure the resulting cell voltages. Complementary demonstrations that may be used to teach students about electrolytic and voltaic cells include "Hoffman Electrolysis" and the "Lemon Battery Contest," respectively. In the demonstration "Basic Electrophoresis," the separation of charged dye molecules in an electric field illustrates the central role of ion migration in electrochemical cells.

Oxidation and Reduction Reactions

While the electrical principles at work in electrochemistry certainly command attention, they should not obscure the chemical reactions that take place. In the experiment "Electrolysis Reactions," students use a simple and inexpensive Petri dish electrolysis set-up to observe the oxidation and reduction reactions of potassium iodide, copper(II) bromide, and sodium chloride in aqueous solution. In "The Tin Man" demonstration, tin(II) ions in a solution of tin(II) chloride are oxidized to tin(IV) ions at the anode and are reduced to tin(0) metal at the cathode. The result is a stunning "tin-man" crystal tree that grows before your eyes!

Quantitative Applications

Electrolysis has many industrial and commercial applications, and the amount of electricity used in electrolysis represents a significant portion of the nation's energy consumption. In the experiment "Quantitative Electrochemistry," students study the relationship between the quantity of electricity used and the amount of product obtained in a copper-electroplating reaction. "Microscale Electrolysis" is a wonderful companion or follow-up demonstration for the "Introduction to Electrochemistry" experiment. The gas mixture generated in the electrolysis of water is collected in a pipet bulb and then ignited with a spark. The resulting "rocket reaction" offers convincing proof that the gas mixture contains the 2:1 stoichiometric ratio of hydrogen and oxygen. Finally, the "Hoffman Electrolysis" demonstration also has an optional quantitative component.

Learning in the Lab

Chemistry is an experimental science! Don't let budget constraints limit what students learn in the lab—electrochemistry does not have to bust the budget. Use the safe, simple, and economical activities in *Electrochemistry* to help you make the most of your resources. The selection of experiments and demonstrations, combined with complete sample data and extensive teacher notes, gives you the ability to design an effective lab curriculum for your students. Best of all, no matter which activities you choose, your students are assured of success. All of the activities in *Electrochemistry* have been thoroughly tested and retested. You know they will work! Use the experiment summaries and concepts on the following pages to locate the concepts you want to teach and to choose experiments and demonstrations that will help you meet your goals.

Format and Features

Flinn ChemTopic™ Labs

All experiments and demonstrations in Flinn ChemTopic™ Labs are printed in a 10⅞″ × 11″ format with a wide 2″ margin on the inside of each page. This reduces the printed area of each page to a standard 8½″ × 11″ format suitable for copying.

The wide margin assures you the entire printed area can be easily reproduced without damaging the binding. The margin also provides a convenient place for teachers to add their own notes.

Concepts — Use these bulleted lists along with state and local standards, lesson plans, and your textbook to identify activities that will allow you to accomplish specific learning goals and objectives.

Background — A balanced source of information for students to understand why they are doing an experiment, what they are doing, and the types of questions the activity is designed to answer. This section is not meant to be exhaustive or to replace the students' textbooks, but rather to identify the core concepts that should be covered before starting the lab.

Experiment Overview — Clearly defines the purpose of each experiment and how students will achieve this goal. Performing an experiment without a purpose is like getting travel directions without knowing your destination. It doesn't work, especially if you run into a roadblock and need to take a detour!

Pre-Lab Questions — Making sure that students are prepared for lab is the single most important element of lab safety. Pre-lab questions introduce new ideas or concepts, review key calculations, and reinforce safety recommendations. The pre-lab questions may be assigned as homework in preparation for lab or they may be used as the basis of a cooperative class activity before lab.

Materials — Lists chemical names, formulas, and amounts for all reagents—along with specific glassware and equipment—needed to perform the experiment as written. The material dispensing area is a main source of student delay, congestion, and accidents. Three dispensing stations per room are optimum for a class of 24 students working in pairs. To safely substitute different items for any of the recommended materials, refer to the *Lab Hints* section in each experiment or demonstration.

Safety Precautions — Instruct and warn students of the hazards associated with the materials or procedure and give specific recommendations and precautions to protect students from these hazards. Please review this section with students before beginning each experiment.

Procedure — This section contains a stepwise, easy-to-follow procedure, where each step generally refers to one action item. Contains reminders about safety and recording data where appropriate. For inquiry-based experiments the procedure may restate the experiment objective and give general guidelines for accomplishing this goal.

Data Tables — Data tables are included for each experiment and are referred to in the procedure. These are provided for convenience and to teach students the importance of keeping their data organized in order to analyze it. To encourage more student involvement, many teachers prefer to have students prepare their own data tables. This is an excellent pre-lab preparation activity—it ensures that students have read the procedure and are prepared for lab.

Post-Lab Questions or Data Analysis — This section takes students step-by-step through what they did, what they observed, and what it means. Meaningful questions encourage analysis and promote critical thinking skills. Where students need to perform calculations or graph data to analyze the results, these steps are also laid out sequentially.

Format and Features
Teacher's Notes

Master Materials List — Lists the chemicals, glassware, and equipment needed to perform the experiment. All amounts have been calculated for a class of 30 students working in pairs. For smaller or larger class sizes or different working group sizes, please adjust the amounts proportionately.

Preparation of Solutions — Calculations and procedures are given for preparing all solutions, based on a class size of 30 students working in pairs. With the exception of particularly hazardous materials, the solution amounts generally include 10% extra to account for spillage and waste. Solution volumes may be rounded to convenient glassware sizes (100-mL, 250-mL, 500-mL, etc.).

Safety Precautions — Repeats the safety precautions given to the students and includes more detailed information relating to safety and handling of chemicals and glassware. Refers to Material Safety Data Sheets that should be available for all chemicals used in the laboratory.

Disposal — Refers to the current *Flinn Scientific Catalog/Reference Manual* for general guidelines and specific procedures governing the disposal of laboratory waste. Because we recommend that teachers review local regulations before beginning any disposal procedure, the information given in this section is for general reference purposes only. However, if a disposal step is included as part of the experimental procedure itself, then the specific solutions needed for disposal are described in this section.

Lab Hints — This section reveals common sources of student errors and misconceptions and where students are likely to need help. Identifies the recommended length of time needed to perform each experiment, suggests alternative chemicals and equipment that may be used, and reminds teachers about new techniques (filtration, pipeting, etc.) that should be reviewed prior to lab.

Teaching Tips — This section puts the experiment in perspective so that teachers can judge in more detail how and where a particular experiment will fit into their curriculum. Identifies the working assumptions about what students need to know in order to perform the experiment and answer the questions. Highlights historical background and applications-oriented information that may be of interest to students.

Sample Data — Complete, actual sample data obtained by performing the experiment exactly as written is included for each experiment. Student data will vary.

Answers to All Questions — Representative or typical answers to all questions. Includes sample calculations and graphs for all data analysis questions. Information of special interest to teachers only in this section is identified by the heading "Note to the teacher." Student answers will vary.

Look for these icons in the *Experiment Summaries and Concepts* section and in the *Teacher's Notes* of individual experiments to identify inquiry-, microscale-, and technology-based experiments, respectively.

Experiment Summaries and Concepts

Experiment

Introduction to Electrochemistry—Electrolysis of Water

Build a simple electrochemical cell to introduce the basic principles of electrochemistry. All you will need are mechanical pencil leads, a 9-V battery, and a U-tube. The purpose of this experiment is to investigate the chemical reaction that takes place when an electric current is forced through water. Compare the amount of gas and indicator color changes at each electrode to identify the oxidation and reduction half-reactions and to determine the overall reaction. Safe and economical, this introductory-level experiment is also a great "classifying matter" activity!

Measuring Cell Potentials—Standard Reduction Potentials

In a voltaic cell, the flow of electrons accompanying a spontaneous oxidation–reduction reaction occurs via an external pathway, and an electric current is produced. What factors determine the ability of a voltaic cell to produce electricity? In this unique microscale version of a classic experiment, students measure the voltage produced by micro-voltaic cells consisting of metals and metal ion solutions on a piece of filter paper. Use the results to calculate the standard reduction potential for each metal and to rank the metals from most active to least active.

Quantitative Electrochemistry—Coulombs, Electrons, and Moles

Principles of electricity and chemistry overlap in electrochemistry, the study of the interconversion of electrical and chemical energy in chemical reactions. What is the relationship between the quantity of electricity and the extent of a chemical reaction in an electrochemical process? The purpose of this experiment is to measure the mass of copper obtained in an electroplating reaction and to relate the amount of product to the amount of electricity that is used.

Electrolysis Reactions—Oxidation and Reduction

When an electric current is passed through an aqueous solution containing sodium sulfate (an electrolyte), the water molecules decompose to give hydrogen gas and oxygen gas. What happens if the electrolyte contains ions that are more easily oxidized or more easily reduced than water molecules? Use a simple and inexpensive "Petri dish" electrolysis set-up to investigate the oxidation and reduction reactions of potassium iodide, copper(II) bromide, and sodium chloride.

Concepts

- Electrochemistry
- Oxidation–reduction
- Electrolysis
- Anode vs. cathode

- Oxidation–reduction
- Voltaic cell
- Standard reduction potential
- Metal activity

- Electrolysis
- Current (amperes)
- Electrical charge (coulombs)
- Faraday constant

- Electrolysis
- Oxidation–reduction
- Anode vs. cathode
- Cell potential

Experiment Summaries and Concepts

Demonstration

Microscale Electrolysis—The Rocket Reaction

When electricity is passed through a solution of water containing an electrolyte, hydrogen gas and oxygen gas are obtained. In this microscale demonstration, the gas mixture is collected in a pipet bulb by water displacement and then ignited with a spark. The resulting "bulb rocket" shoots across the room, proving that the electrolysis gas mixture contains the 2:1 stoichiometric ratio of hydrogen and oxygen.

Hoffman Electrolysis—Color Enhanced!

Use the Hoffman apparatus—a great demonstration device—to illustrate the basic principles of electrochemistry. The presence of an indicator adds color and helps students see what's going on. What reaction occurs at the anode versus the cathode? What ions are produced at each electrode? Quantitative principles of electrochemistry can also be demonstrated with the proper equipment.

The Tin Man—Tin(IV), Tin(II), and Tin(0)

Grow a beautiful "tin-man" crystal tree by running an electric current through a solution of tin(II) chloride! Tin(II) ions are oxidized to tin(IV) ions at the anode and are reduced to tin(0) metal at the cathode.

Orange Juice Clock—Electricity from a Chemical Reaction

Students are always looking at the clock anyway, but this demonstration will really get them looking—and wondering! Build an electrochemical cell using orange juice and copper and magnesium metal electrodes. The resulting spontaneous oxidation–reduction reaction generates enough electricity to run a battery-powered clock. Use this demonstration to jump-start your lesson plans!

Basic Electrophoresis—Migration of Ions in an Electric Field

One of the most poorly understood concepts in electrochemistry has to do with the flow of electricity in an electrochemical cell. Electrons carry electricity through the external circuit, but ions carry the current through the solution. A central feature of electrochemical cells, ion migration in an electric field is also the basic principle in electrophoresis. Watch charged dye molecules move in opposite directions in this quick and easy introduction to electrophoresis.

Lemon Battery Contest—Metal Activity and Cell Potentials

Design your own battery using strips of different metals and a lemon or an orange. Good luck! What is the largest voltage that can be obtained in a lemon battery? Use the results obtained with different metals to deduce the reaction that takes place in a lemon battery.

Concepts

- Electrolysis
- Oxidation–reduction
- Decomposition reaction
- Combustion reaction

- Electrolysis
- Oxidation–reduction
- Anode vs. cathode
- Current, coulombs, electrons

- Electrolysis
- Oxidation–reduction
- Anode vs. cathode

- Electrochemistry
- Cell potential
- Metal activity
- Anode vs. cathode

- Electrophoresis
- Migration of ions
- Positive and negative electrodes

- Electrochemistry
- Cell potential
- Metal activity
- Anode vs. cathode

Page 1 – **Introduction to Electrochemistry**

Teacher Notes

Introduction to Electrochemistry
Electrolysis of Water

Introduction

Electrochemistry is the study of the relationship between electrical forces and chemical reactions. There are two basic types of electrochemical processes. In a voltaic cell, commonly known as a battery, the chemical energy from a spontaneous oxidation–reduction reaction is converted into electrical energy. In an electrolytic cell, electricity from an external source is used to "force" a nonspontaneous chemical reaction to occur. What chemical reaction will take place when an electric current flows through water?

Concepts

- Electrochemistry
- Electrolysis
- Oxidation–reduction
- Anode vs. cathode

Background

The first electrochemical process to produce electricity was described in 1800 by the Italian scientist Alessandro Volta, a former high school teacher. Acting on the hypothesis that two dissimilar metals could serve as a source of electricity, Volta constructed a stacked pile of alternating silver and zinc plates separated by pads of absorbent material soaked in saltwater. When Volta moistened his fingers and repeatedly touched the top and bottom metal plates, he experienced a series of small electric shocks. The "voltaic pile," as it came to be called, was the first battery—a chemical method of generating an electric current. Within months, William Nicholson and Anthony Carlisle in England attempted to confirm the production of electric charges on the upper and lower plates in a voltaic pile using an electroscope. In order to connect the plates to the electroscope, Nicholson and Carlisle added some water to the uppermost metal plate and inserted a wire to the electroscope. To their surprise, Nicholson and Carlisle observed the formation of a gas, which they identified as hydrogen. Nicholson and Carlisle then filled a small tube with river water and inserted wires from the voltaic pile into each end of the tube. Two different gases were generated, one at each wire— Nicholson and Carlisle had discovered electrolysis.

Experiment Overview

The purpose of this experiment is to investigate the electrolysis of water in an electrochemical cell. Two carbon pencil "leads" will be inserted into the opposite ends of a U-tube containing water, sodium sulfate, and bromthymol blue. An electric current will be passed through the solution by connecting the pencil leads to the positive and negative terminals of a 9-volt battery. (See Figure 1 on page 2.) The pencil leads act as external conductors and provide a surface for the chemical reaction. Sodium sulfate, an ionic compound, is needed to improve the current flow through the solution. Bromthymol blue, an acid–base indicator, will help to identify the changes occurring in the solution as the electrolysis proceeds. Bromthymol blue is yellow in acidic solutions (pH <6.0), blue in basic solution (pH >7.6), and various shades of green at intermediate pH values (pH = 6.0–7.6).

Although electrochemistry is traditionally taught late in the second half of the chemistry course, this introductory-level experiment may be used very early in the curriculum. It is an excellent "classifying matter" activity to prove that water is a compound.

Introduction to Electrochemistry – Page 2

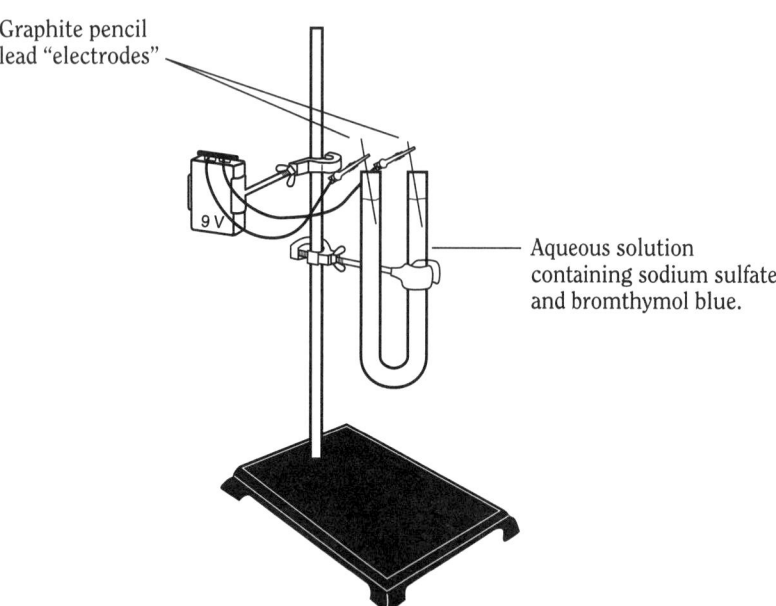

Figure 1. U-Tube Setup for the Electrolysis of Water.

Pre-Lab Questions

Recall that any oxidation–reduction reaction may be written as the sum of two half-reactions, an oxidation half-reaction and a reduction half-reaction. Electrons flow from the substance that is oxidized (which loses electrons), to the substance that is reduced (which gains electrons). If the half-reactions are separated, the electrons will flow through an external conductor rather than through the solution. This is the basis of electrochemistry. In electrolysis, the electron flow is not spontaneous, but rather is "forced" through a battery.

1. A decomposition reaction may be defined as any reaction in which one reactant, a compound, breaks down to give two or more products. Write the balanced chemical equation for the decomposition of water to its elements.

2. (a) Assign oxidation states to the hydrogen and oxygen atoms in each substance in the above chemical equation.

   (b) Based on the changes in oxidation states for each atom, identify the atom that is oxidized and the atom that is reduced in the decomposition of water.

3. Balance the following oxidation and reduction half-reactions for the decomposition of water. *Hint:* Hydrogen ions (H^+) and hydroxide ions (OH^-) are required to balance atoms and charge.

   ☐ H_2O → O_2 + ☐ H^+ + ☐ e^-

   ☐ H_2O + ☐ e^- → H_2 + ☐ OH^-

4. Explain how the oxidation and reduction half-reactions may be combined to give the balanced chemical equation for the decomposition of water. What happens to the electrons and to the H^+ and OH^- ions?

Teacher Notes

Inexpensive microscale U-tubes may be constructed by bending 7-mm glass tubing into a U-shape using a Bunsen burner flame.

*Page 3 – ***Introduction to Electrochemistry***

Teacher Notes

Materials

Bromthymol blue indicator solution, 0.04%, 2 mL

Sodium sulfate solution, Na_2SO_4, 0.5 M, 30 mL

Battery, 9-V

Battery cap with alligator clip leads

Beaker, 50-mL

Beral-type pipet

Carbon pencil "leads," 0.9-mm, 2

Support clamps, 2

Ring (support) stand

U-shaped tube, 150-mm

Safety Precautions

To extend the life of the battery, avoid connecting the positive and negative terminals to each other. Wear chemical splash goggles, chemical-resistant gloves, and a chemical-resistant apron. Wash hands thoroughly with soap and water before leaving the lab.

Procedure

1. Attach the battery cap with alligator clips to the 9-V battery. The battery may be placed in a support clamp, if needed, to prevent tension when the alligator clips are attached to the carbon pencil leads (step 5).

2. Clamp the U-tube to a ring stand for support.

3. Obtain about 30 mL of 0.5 M sodium sulfate solution in a small beaker. Using a Beral-type pipet, add about 2 mL of bromthymol blue indicator and stir the solution to evenly distribute the indicator color. Observe and record the initial indicator color of the solution.

4. Carefully pour the electrolysis is solution into the U-tube until the liquid level is about 2 cm from the top.

5. Connect the pencil leads to the alligator clips on the battery cap and insert one pencil lead into each side of the U-tube.

6. Observe and record all changes as the current flows through the electrolysis solution. Be specific—compare the changes at the pencil lead electrodes attached to the positive (+) and negative (–) terminals of the battery.

7. Allow the current to flow through the solution for about 5 minutes.

8. Remove the pencil leads from the solution and disconnect the alligator clips.

9. Disconnect the battery cap from the battery and return both to their proper location.

10. Pour the solution from the U-tube into a small beaker. Observe and record the final indicator color of the mixed solution.

The sodium sulfate electrolysis solution may be reused by several classes during the day. Students will not need to add the indicator separately if the electrolysis solution is being recycled. The color changes observed at each electrode can be reversed by switching the electrodes after students have completed their initial observations (step 7). This may help students connect the "sign" of the electrode with the reaction taking place.

Name: _____

Class/Lab Period: _____

Introduction to Electrochemistry

Data Table

Initial Indicator Color of the Electrolysis Solution	
Changes Occurring at the Positive (+) Electrode	
Changes Occurring at the Negative (−) Electrode	
Final Indicator Color of the Mixed Electrolysis Solution	

Post-Lab Questions *(Use a separate sheet of paper to answer the following questions.)*

1. Suggest an explanation for the initial indicator color of the electrolysis solution.

2. Describe at least three observations that indicate a chemical reaction has occurred during the electrolysis of water.

3. What are the two functions of the pencil lead electrodes?

4. (a) Compare the color changes observed at the positive (+) and negative (−) electrodes. What ions were produced at each electrode?

    (b) Write out the oxidation and reduction half-reactions for the decomposition of water and identify which reaction occurred at each electrode, based on the indicator color changes.

5. Compare the rates of gas evolution at the positive (+) and negative (−) electrodes. What gas was produced at each electrode? Explain, based on the balanced chemical equation for the decomposition of water. (See *Pre-Lab Question #1*.)

6. Suggest an explanation for the *final* indicator color of the mixed electrolysis solution (step 10 in the *Procedure*).

7. Think about the flow of electrons and current in the electrolysis of water. What do the positive and negative signs on a battery signify?

8. *(Optional)* Decomposing water to its elements requires energy in the form of electricity. The reverse process, combining hydrogen and oxygen to form water, may be used to generate electricity in a *fuel cell*. Research and describe the basic features of a fuel cell.

Teacher Notes

LEO the lion goes GER! Students love this mnemonic device for remembering the definitions of oxidation and reduction.

Loss of **E**lectrons = **O**xidation (LEO)

Gain of **E**lectrons = **R**eduction (GER)

Teacher's Notes

Teacher's Notes
Introduction to Electrochemistry

Master Materials List *(for a class of 30 students working in pairs)*

Bromthymol blue indicator solution, 0.04%, 50 mL
Sodium sulfate solution, Na_2SO_4, 0.5 M, 500 mL

Batteries, 9-V, 15	Carbon pencil "leads," 0.9-mm, 30
Battery caps with alligator clip leads, 15	Support clamps, 30
Beakers, 50-mL, 15	Ring (support) stands, 15
Beral-type pipets, 15	U-shaped tubes, 150-mm, 15

Preparation of Solutions *(for a class of 30 students working in pairs)*

Bromthymol Blue Indicator Solution, 0.04% Aqueous: Dissolve 0.04 g of bromthymol blue indicator in about 50 mL of distilled or deionized water. Stir to dissolve and then dilute to 100 mL with water.

Sodium Sulfate, 0.5 M: Dissolve 161 g of sodium sulfate decahydrate ($Na_2SO_4·10H_2O$) in about 500 mL of distilled or deionized water. Stir to dissolve and then dilute to 1 L with water.

Safety Precautions

To extend the life of the battery, avoid touching the positive and negative terminals to each other. Wear chemical splash goggles and chemical-resistant gloves. Please review current Material Safety Data Sheets for additional safety, handling, and disposal information. Remind students to wash their hands thoroughly with soap and water before leaving the lab.

Disposal

Please consult your current *Flinn Scientific Catalog/Reference Manual* for general guidelines and specific procedures governing the disposal of laboratory waste. The electrolysis solution may be disposed of down the drain with plenty of excess water according to Flinn Suggested Disposal Method #26b.

Lab Hints

- The laboratory work for this experiment can easily be completed in 20–30 minutes. The remaining time in a typical 50-minute lab period may be used to carry out some complementary demonstrations. See the *Teaching Tips*.

- Sodium sulfate is used as a source of dissolved ions to increase the current flow through the solution. In the absence of added electrolyte, no reaction will occur when the battery is connected to the pencil leads—there are no ions to "carry" the current through the solution. The rate of electrolysis increases as the concentration of sodium sulfate increases (compare 0.1 M and 1 M solutions). The conductivity of a sodium sulfate solution versus pure water may be demonstrated using a conductivity tester (available from Flinn Scientific, Catalog No. AP5355).

Teacher's Notes

- The sulfate ion is an extremely weak base (the pK_a for its conjugate acid, HSO_4^-, is 2.0). The initial indicator color for the electrolysis solution may be more blue-green rather than green. According to the *Merck Index,* the pH of a sodium sulfate solution is 6.0–7.6. Test a small amount of the sodium sulfate stock solution with bromthymol blue indicator before class—the solution should turn green. If the solution is blue, add one drop of 1 M hydrochloric acid to the stock solution. If the solution is yellow, add one drop of 1 M sodium hydroxide to the stock solution.

- The sodium sulfate electrolysis solution may be reused by several classes during the day. Recycling the solution in this way means the indicator will already be present—instruct students to omit step 3 in the *Procedure*. Adjust the pH of the electrolysis solution as described above, if necessary.

- The odor of chlorine may be observed at the anode if sodium chloride is substituted for sodium sulfate as the electrolyte in the electrolysis solution. Based on their standard reduction potentials, oxidation of chloride ion to chlorine ($E°$ = –1.36 V) is less favorable than oxidation of water to oxygen ($E°$ = –1.23 V). However, there is a significant overvoltage for oxidation of water, and oxidation of chloride competes with oxidation of water under typical electrolysis conditions. Although the cause of the overvoltage is poorly understood, it is generally believed to be due to a kinetically slow reaction at the anode. See the experiment "Electrolysis Reactions" for additional examples of the electrolysis of salt solutions.

Teaching Tips

- See "Hoffman Electrolysis" in the *Demonstrations* section of this book for a large, demonstration-scale version of this experiment using the Hoffman electrolysis apparatus. Using the Hoffman apparatus makes it possible to measure the volume of gas generated at each electrode, collect the gases, and test their properties. If the time and current are measured, the amount of hydrogen gas collected may also be used for quantitative calculations of the Faraday constant.

- In the demonstration "Microscale Electrolysis—The Rocket Reaction," the gas mixture generated in the microscale electrolysis of water is collected in a jumbo pipet bulb. Igniting the gas mixture with a piezoelectric igniter propels the resulting "bulb rocket" across the room—a great way to compare the properties of a 2:1 stoichiometric mixture of hydrogen and oxygen versus pure hydrogen.

- The discovery of current electricity by Alessandro Volta in 1800 led to the almost immediate discovery of electrolysis, which led, in turn, to the rapid discovery of new chemical elements. Humphry Davy, a professor at the Royal Institution in London, began extensive studies in electrochemistry that culminated in 1807–1808 with his discoveries of the metals potassium, sodium, magnesium, calcium, strontium, and barium.

- See the following Web site for information about the principles and design of fuel cells. http://www.netl.doe.gov/coolscience/teacher/lesson-plans/lesson 6.pdf (accessed October 2004).

Teacher's Notes

Teacher Notes

Answers to Pre-Lab Questions *(Student answers will vary.)*

1. A decomposition reaction may be defined as any reaction in which one reactant, a compound, breaks down to give two or more products. Write the balanced chemical equation for the decomposition of water to its elements.

    $$2H_2O(l) \xrightarrow{electricity} 2H_2(g) + O_2(g)$$

2. (a) Assign oxidation states to the hydrogen and oxygen atoms in each substance in the above chemical equation.

    Oxidation state of H in H_2O = +1. Oxidation state of O atom in H_2O = –2.

    Oxidation state of H and O in elemental H_2 and O_2 = 0.

   (b) Based on the changes in oxidation states for each atom, identify the atom that is oxidized and the atom that is reduced in the decomposition of water.

    The oxygen atom is oxidized (oxidation state increases from –2 to 0). The hydrogen atom is reduced (oxidation state decreases from +1 to 0).

3. Balance the following oxidation and reduction half-reactions for the decomposition of water. *Hint:* Hydrogen ions (H^+) and hydroxide ions (OH^-) are required to balance atoms and charge.

    $\boxed{2}\ H_2O \rightarrow O_2 + \boxed{4}\ H^+ + \boxed{4}\ e^-$

    $\boxed{2}\ H_2O + \boxed{2}\ e^- \rightarrow H_2 + \boxed{2}\ OH^-$

4. Explain how the oxidation and reduction half-reactions may be combined to give the balanced chemical equation for the decomposition of water. What happens to the electrons and to the H^+ and OH^- ions?

    The electrons must balance or "cancel out" when the oxidation and reduction half-reactions are combined. The reduction half-reaction must therefore be multiplied by a factor of two. H^+ and OH^- ions generated in the individual half-reactions combine to form water molecules.

Teacher's Notes

Sample Data

Student data will vary.

Data Table

Initial Indicator Color of the Electrolysis Solution	Green
Changes Occurring at the Positive (+) Electrode	Gas bubbling observed at the surface of the carbon electrode. Gas appears to dissolve in solution. Color of solution changes to yellow.
Changes Occurring at the Negative (–) Electrode	Rapid gas bubbling observed at electrode surface. Gas bubbles are large and do not dissolve. Color of solution changes to blue.
Final Indicator Color of the Mixed Electrolysis Solution	Green

Answers to Post-Lab Questions *(Student answers will vary.)*

1. Suggest an explanation for the initial indicator color of the electrolysis solution.

    The initial indicator color of the electrolysis solution was green—the solution is near neutral, pH = 6.0–7.6. This is consistent with the fact that sodium sulfate is a neutral salt (like sodium chloride). The resulting solution is neither acidic nor basic.

2. Describe at least three observations that indicate a chemical reaction has occurred during the electrolysis of water.

    Signs of a chemical reaction: Formation of gas bubbles at each electrode; indicator color change to yellow at the positive electrode; indicator color change to blue at negative electrode. **Note to teachers:** *The gas bubbles appear visibly different at the cathode versus the anode.*

3. What are the two functions of the pencil lead electrodes?

    The electrodes act as external conductors for the electric current between the battery and the solution and also provide a surface for the chemical reactions.

4. (a) Compare the color changes observed at the positive (+) and negative (–) electrodes. What ions were produced at each electrode?

    The indicator color changed to yellow at the (+) electrode. This is due to the formation of H^+ (H_3O^+) ions. Bromthymol blue is yellow in acidic solutions (pH <6), when the concentration of H^+ ions is greater than the concentration of OH^- ions. The indicator color changed to blue at the negative electrode. This is due to the formation of OH^- ions. Bromthymol blue is blue in basic solutions (pH >7.6), when the concentration of OH^- ions is greater than the concentration of H^+ ions.

Teacher's Notes

Teacher Notes

4. (b) Write out the oxidation and reduction half-reactions for the decomposition of water and identify which reaction occurred at each electrode, based on the indicator color changes.

$$2H_2O(l) \rightarrow 4H^+(aq) + O_2(g) + 4e^-$$ Oxidation occurred at the (+) electrode. (Production of H^+ ions)

$$2H_2O(l) + 2e^- \rightarrow H_2(g) + 2OH^-(aq)$$ Reduction occurred at the (–) electrode. (Production of OH^- ions)

5. Compare the rates of gas evolution at the positive (+) and negative (–) electrodes. What gas was produced at each electrode? Explain, based on the balanced chemical equation for the decomposition of water. (See *Pre-Lab Question #1*.)

The rate of gas evolution was greater at the negative electrode, where hydrogen gas is formed. According to the balanced chemical equation for the decomposition of water, two moles of hydrogen gas are formed for every mole of oxygen gas that is released.

6. Suggest an explanation for the *final* indicator color of the mixed electrolysis solution.

The final indicator color of the mixed electrolysis solution was green (neutral)—the total number of H^+ ions produced at the positive electrode is equal to the total number of OH^- ions produced at the negative electrode. The balanced chemical equation for the overall reaction at both electrodes shows there is no net excess of either ion. (The ions combine to produce water molecules.)

7. Think about the flow of electrons and current in the electrolysis of water. What do the positive and negative signs on a battery signify?

Electrons flow from the negative battery terminal to the negative electrode, where they are "consumed" in the reduction half-reaction. Oxidation occurs at the positive electrode, where electrons are released and flow into the positive terminal on the battery. The battery acts as an "electron pump," pushing electrons into one electrode and pulling them from the other electrode. Electric current flows through the electrolysis solution via the migration of ions.

Note to teachers: *In all types of electrochemical cells, electrons carry the current through the external wire, while ions carry the current through the solution. Anions move toward the anode, cations move toward the cathode.*

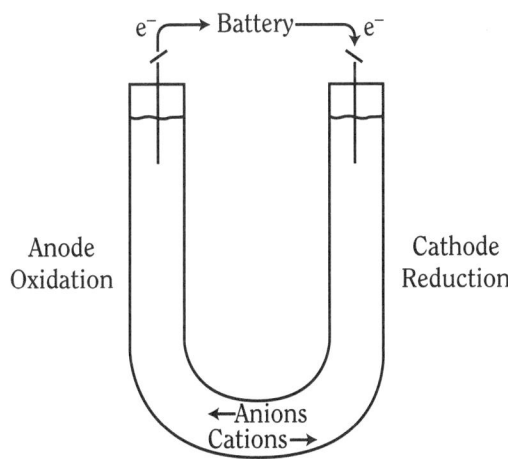

Introduction to Electrochemistry

Teacher's Notes

8. *(Optional)* Decomposing water to its elements requires energy in the form of electricity. The reverse process, combining hydrogen and oxygen to form water, may be used to generate electricity in a *fuel cell*. Research and describe the basic features of a fuel cell.

    A fuel cell is a device for the direct production of electricity from the energy released in a chemical reaction. Although in theory many different fuels may be used as the energy source, most of the research and engineering today refers to hydrogen fuel cells, in which hydrogen is the energy source. The chemical reaction is the combustion (combination) reaction of hydrogen with oxygen to produce water and energy.

Teacher Notes

Measuring Cell Potentials
Standard Reduction Potentials

Introduction

In an oxidation–reduction reaction, electrons flow from the substance that is oxidized, which loses electrons, to the substance that is reduced, which gains electrons. In a voltaic cell, the flow of electrons accompanying a spontaneous oxidation–reduction reaction occurs via an external pathway, and an electric current is produced. What factors determine the ability of a voltaic cell to produce electricity?

Concepts

- Oxidation–reduction
- Standard reduction potential
- Voltaic cell
- Activity series of metals

Background

The basic design of a *voltaic cell* is shown in Figure 1 for the net reaction of zinc and hydrochloric acid. The substances involved in each half-reaction are separated into two compartments connected by an external wire and a salt bridge.

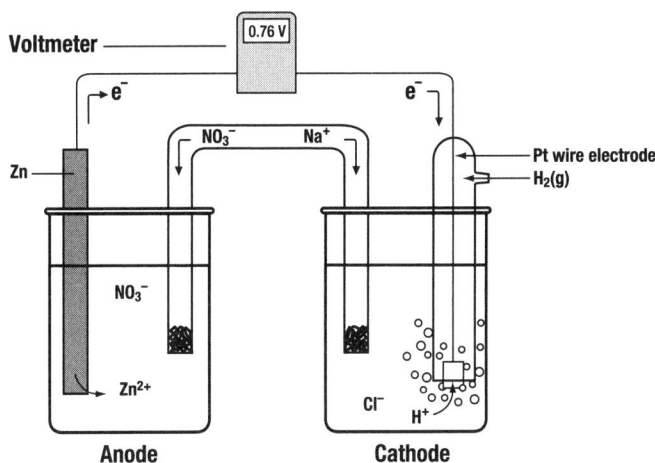

$Zn(s) \rightarrow Zn^{2+}(aq) + 2e^-$ $2H^+(aq) + 2e^- \rightarrow H_2(g)$

Figure 1. Components of an Electrochemical Cell.
Net Reaction: $Zn(s) + 2H^+(aq) \rightarrow Zn^{2+}(aq) + H_2(g)$

Each half-reaction takes place at the surface of a metal plate or wire called an *electrode*. The electrode at which oxidation occurs is called the *anode*, while the electrode at which reduction occurs is called the *cathode*. Electrons flow spontaneously from the anode (the negative electrode) to the cathode (the positive electrode). Charge buildup at the electrodes is neutralized by connecting the half-cells internally by means of a *salt bridge*, a porous barrier containing sodium nitrate or another electrolyte. Dissolved ions flow through the salt bridge to either electrode, thus completing the electrical circuit.

The ability of a voltaic cell to produce an electric current is called the *cell potential* and is measured in volts. If the cell potential is large, there is a large "electromotive force" pushing

The cell potential may be described as resulting from an electron "tug of war." In the cell shown in Figure 1, H^+ ions have a greater tendency than Zn^{2+} ions to "pull" electrons toward them.

Measuring Cell Potentials – Page 2

or pulling electrons through the circuit from the anode to the cathode. The cell potential for a spontaneous chemical reaction in a voltaic cell is always positive. The *standard cell potential* ($E°_{cell}$) is defined as the maximum potential difference between the electrodes of an electrochemical cell under standard conditions—25 °C, 1 M concentrations of ions, and 1 atm pressure (for gases).

It is impossible to directly measure the potential for a single electrode. The overall cell potential for an electrochemical cell may be expressed, however, as the difference between the *standard reduction potentials* ($E°_{red}$) for the reactions at the cathode and at the anode (Equation 1).

$$E°_{cell} = E°_{red} \text{ (cathode)} - E°_{red} \text{ (anode)} \qquad \text{Equation 1}$$

The standard reduction potential is defined as the voltage that a *reduction* half-cell will develop under standard conditions when it is combined with the *standard hydrogen electrode* (SHE), which is arbitrarily assigned a potential of zero volts (Equation 2).

$$2H^+(aq, 1\,M) + 2e^- \rightarrow H_2(g, 1\,atm) \qquad \text{Equation 2}$$
$$E°_{red} \text{ (SHE)} = 0$$

For the zinc/hydrochloric acid voltaic cell shown in Figure 1, the measured cell potential is equal to 0.76 V. Substituting this value and the zero potential for the SHE into Equation 1 gives a value of –0.76 V for the standard reduction potential of the Zn^{2+}/Zn half-cell.

$$E°_{red} \text{ (cathode)} - E°_{red} \text{ (anode)} = E°_{cell}$$
$$E°_{red} \text{ (SHE)} - E°_{red} (Zn^{2+}/Zn) = 0.76\,V$$
$$0 - E°_{red} (Zn^{2+}/Zn) = 0.76\,V$$
$$E°_{red} (Zn^{2+}/Zn) = -0.76\,V$$

When two half-cells are combined in a voltaic cell, the reaction that has a more positive standard reduction potential will occur as a reduction, while the reaction that has a less positive (or negative) standard reduction potential will be reversed and will take place as an oxidation. In this experiment, electrochemical cells consisting of different metal ion/metal half-cells, e.g., copper(II) sulfate/copper metal versus zinc sulfate/zinc metal, will be tested. The "direction" of each reaction—the identity of the anode and the cathode—will be determined when a positive voltage is observed. A positive voltage means that the half-cells have been properly connected to the positive and negative leads on the voltmeter so that a spontaneous reaction will occur. Recall that in a voltaic cell, the positive electrode is the cathode (the site of reduction) and the negative electrode is the anode (the site of oxidation).

Experiment Overview

The purpose of this experiment is to measure cell potentials for a series of micro-voltaic cells. Individual half-cells will be constructed by placing a small piece of metal onto 1–2 drops of its metal ion solution on a piece of filter paper. A "salt bridge" between the half-cells will be provided by placing a few drops of aqueous sodium nitrate on the filter paper along the path connecting the half-cells. Voltages will be measured using a multimeter or a voltage probe attached to a computer interface.

Teacher Notes

The cell potential is defined as the maximum voltage generated by a cell under standard conditions, when no current is being drawn from the cell. (Any current flow will dissipate some energy as heat and decrease the voltage.) Digital voltmeters are designed to draw negligible current. Conventional analog voltmeters will not give accurate cell potentials.

Page 3 – **Measuring Cell Potentials**

> Teacher Notes

Pre-Lab Questions

Standard reduction potentials ($E°_{red}$) are defined for *reduction* half-reactions. For a voltaic cell, $E°_{red}$ for the cathode (reduction) must be more positive than $E°_{red}$ for the anode (oxidation). Consider the following two metals and their standard reduction potentials.

$$Co^{2+}(aq) + 2e^- \rightarrow Co(s) \qquad E°_{red} = -0.28 \text{ V}$$

$$Ag^+(aq) + e^- \rightarrow Ag(s) \qquad E°_{red} = 0.80 \text{ V}$$

1. If half-cells for these two metals were combined in a voltaic cell, what reaction would take place at the anode? What reaction would take place at the cathode?

2. Calculate the cell potential for the overall (spontaneous) reaction under standard conditions.

3. Write the balanced chemical equation for the overall reaction.

4. The activity of a metal is defined as its ease of oxidation (an active metal is one that is easily oxidized). Based on their standard reduction potentials, which is the more active metal, cobalt or silver?

Materials

Copper strips or foil, Cu, 1-cm²	Beral-type pipets or eyedroppers, 6
Copper(II) sulfate solution, $CuSO_4$, 1 M, 1 mL	Computer interface system (LabPro)
Iron strips or sheet, Fe, 1-cm²	Data collection software (LoggerPro)
Iron(II) sulfate solution, $FeSO_4$, 1 M, 1 mL	Filter paper, quantitative, 9-cm
Magnesium ribbon, Mg, 1 cm	Multimeter or Voltage probe
Magnesium sulfate solution, $MgSO_4$, 1 M, 1 mL	Petri dish or acetate transparency
Silver foil, Ag, 1-cm²	Sandpaper or steel wool
Silver nitrate solution, $AgNO_3$, 1 M, 1 mL	Scissors
Sodium nitrate solution, $NaNO_3$, 1 M, 2 mL	Tweezers or forceps
Zinc strips or foil, Zn, 1-cm²	Wash bottle and distilled water
Zinc sulfate solution, $ZnSO_4$, 1 M, 1 mL	White paper and pencil

Safety Precautions

Silver nitrate solution is a corrosive liquid and toxic by ingestion. It will stain skin and clothing. Copper(II) sulfate solution is toxic by ingestion. Iron(II) sulfate and zinc sulfate solutions are slightly toxic. Magnesium metal is a flammable solid; avoid contact with flames and heat. Metal pieces may have sharp edges—handle with care. Avoid contact of all chemicals with eyes and skin. Wear chemical splash goggles, chemical-resistant gloves, and a chemical-resistant apron. Wash hands thoroughly with soap and water before leaving the lab.

Larger diameter filter paper (11 cm is most common) may be cut down in size to fit in the Petri dish. If the filter paper will be placed on acetate transparency or plastic wrap (see step 5 in the Procedure*), the size of the filter paper does not matter.*

Measuring Cell Potentials – *Page 4*

Procedure

Teacher Notes

Part A. Cell Potentials versus Zinc as the Reference Electrode

1. Label a sheet of paper with the names of the five metals to be tested (copper, iron, magnesium, silver, and zinc). Obtain one small piece of each metal and place it on the paper.

2. Polish the metal pieces with sandpaper or steel wool, if necessary, to obtain fresh, shiny surfaces. Wipe the metal strips clean with paper towels to remove any bits of steel wool adhering to the metal.

3. Obtain a piece of filter paper. Using a *pencil*, draw five small circles in a symmetrical pattern on the filter paper, and connect the circles by means of a path of dots, as shown in Figure 2.

4. Using a pair of scissors, cut wedges between the circles and remove the wedges. Label the circles Cu, Fe, Mg, Ag, and Zn. See Figure 3.

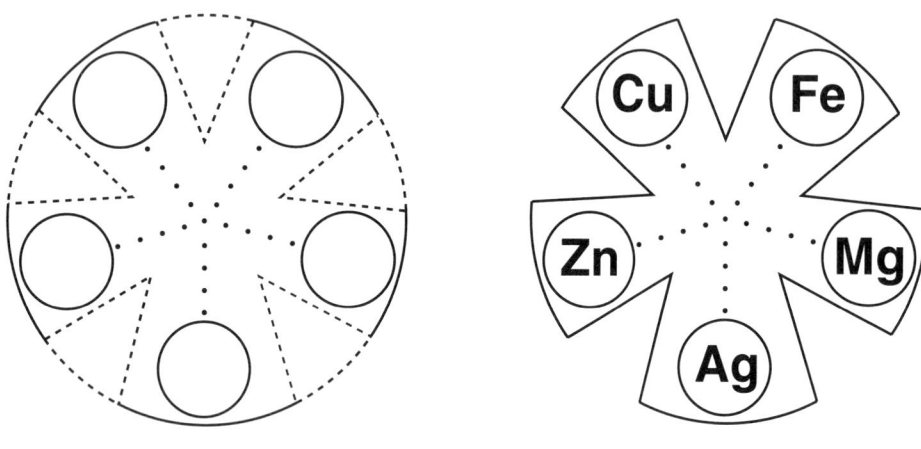

Figure 2.   **Figure 3.**

5. Place the labeled filter paper in a Petri dish or on a sheet of acetate transparency or plastic wrap.

6. Using a separate Beral-type pipet or eyedropper for each solution, place 1–2 drops of metal ion solution onto its corresponding circle on the filter paper (copper(II) sulfate on Cu, iron(II) sulfate on Fe, etc.) *Note:* Add more drops of metal ion solution to each circle as needed during the course of the experiment when the paper dries.

7. Using forceps, place each metal on the "wet spot" in the appropriate circle on the filter paper. Wipe the forceps clean with a paper towel for each metal to avoid contaminating the metals with other metal ions.

8. Using a clean, Beral-type pipet, add several drops of sodium nitrate solution all along the path of dots connecting the metals. Be sure there is a continuous trail of sodium nitrate solution between each circle and the center. *Note:* It may be necessary to add more sodium nitrate along the path of dots as the paper dries during the course of the experiment.

Note: If you will be using a multimeter for voltage measurements, skip steps 9–12 and proceed directly to step 13.

Flinn ChemTopic™ Labs — Electrochemistry

Teacher Notes

9. *(Optional)* Connect the interface system to the computer or calculator and plug the voltage probe into the interface.

10. *(Optional)* Select *Show Sensors* under "Experiment" on the main screen in the data collection software. Choose the correct sensor for the channel in which the voltage probe is connected.

11. *(Optional)* Select *Data Collection* under "Experiment," followed by *Selected Events* or *Events with Entry*.

12. *(Optional)* Press *Collect* on the main screen to begin voltage measurements.

13. Zinc metal ($E°_{red}$ = –0.76 V will be used as the "reference electrode" in Part A. Place the positive lead from the multimeter or voltage probe on the piece of zinc and the negative lead on the piece of copper metal. Read the live voltage. *If the voltage drops to 0.00 V or if a negative voltage is displayed, reverse the leads—place the negative lead on the zinc and the positive lead on the copper.*

14. In the Data Table, record which metal is the positive electrode (cathode) and which metal is the negative electrode (anode) when a positive voltage is obtained.

15. When the voltage reading stabilizes, record the positive voltage in the Data Table. *(Optional)* Press *Keep* on the main screen to automatically record the voltage measurement using the voltage probe.

16. Repeat steps 13–15 three times to measure the cell potentials for the other metals (iron, magnesium, and silver) versus zinc as the reference electrode. Remember to record which metal is the (+) electrode and which metal is the (–) electrode when a positive voltage is obtained. Replenish the metal ion solutions or the salt bridge (sodium nitrate) solution as needed during the course of the experiment if the paper dries out.

17. Using Equation 1 in the *Background* section, calculate the experimental value of the standard reduction potential ($E°_{red}$) for each metal and record the result in the Data Table. Recall that $E°_{red}$ for zinc is equal to –0.76 V.

Part B. Predicted and Measured Cell Potentials

18. Select *two metals* other than zinc (e.g., copper and magnesium) and use their experimental $E°_{red}$ values to *predict* the cell potential for a voltaic cell made up of these two half-cells. Show the calculation and record the predicted value of $E°_{cell}$ in the Data Table. Note which metal should be the anode and the cathode.

19. Measure the cell potential for the voltaic cell and record the value in the Data Table.

20. Repeat steps 18 and 19 for another combination of two metals, not including zinc.
    Note: There are six possible metal combinations (voltaic cells) in Part B. The teacher may assign different voltaic cells to different groups so that all the data will be collected. Class results may then be shared.

It is worth repeating the note in step 13 for emphasis. The voltage must be positive and it is the students' responsibility to place the voltage leads correctly. Take a class "time-out" after Part A to illustrate sample calculations for steps 17 and 18.

Name: _____

Class/Lab Period: _____

Measuring Cell Potentials

Data Table

Part A. Cell Potentials versus Zinc as the Reference Electrode

Metal (M)	Positive Electrode (Cathode)	Negative Electrode (Anode)	Measured Cell Potential $E°_{cell}$	Calculated Value $E°_{red}$ (M)
Copper				
Iron				
Magnesium				
Silver				

Part B. Predicted and Measured Cell Potentials

Cathode/Anode	Calculation (Equation 1)	Predicted Cell Potential	Measured Cell Potential

Post-Lab Questions *(Use a separate sheet of paper to answer the following questions.)*

1. Write the balanced chemical equation for the overall reaction in each voltaic cell in Part A.

2. Consider the class results for all possible cell combinations in Parts A and B: Which metal was most easily oxidized (it always appeared as the anode)? Which metal ions were most easily reduced (the corresponding metal always appeared as the cathode)?

3. Rank the five metals tested (including zinc) from most positive to most negative standard reduction potential. Write a general statement describing the relationship between the standard reduction potential of a metal and metal activity.

4. Look up the literature values of the standard reduction potentials for Cu, Fe, Mg, and Ag and calculate the percent error for each. *Hint:* Note the symbol for "absolute value."

$$\text{Percent error} = \frac{|\text{Experimental value} - \text{Literature value}|}{\text{Literature value}} \times 100\%$$

Teacher's Notes
Measuring Cell Potentials

Master Materials List *(for a class of 30 students working in pairs)*

Copper strips or foil, Cu, 1-cm², 15	Beral-type pipets or eyedroppers, 90*
Copper(II) sulfate solution, CuSO₄, 1 M, 15 mL	Computer interface systems (LabPro), 15
Iron strips or sheet, Fe, 1-cm², 15	Data collection software (LoggerPro)
Iron(II) sulfate solution, FeSO₄, 1 M, 15 mL	Filter paper, quantitative, 9-cm, 15 sheets
Magnesium ribbon, Mg, 1-cm pieces, 15	Multimeters or voltage probes, 15
Magnesium sulfate solution, MgSO₄, 1 M, 15 mL	Petri dishes or acetate transparency, 15
Silver foil, Ag, 1-cm², 15	Sandpaper or Steel wool, 15 pieces
Silver nitrate solution, AgNO₃, 1 M, 15 mL	Scissors, 15
Sodium nitrate solution, NaNO₃, 1 M, 30 mL	Tweezers or forceps, 15
Zinc strips or foil, Zn, 1-cm², 15	Wash bottles and distilled water, 15
Zinc sulfate solution, ZnSO₄, 1 M, 15 mL	White paper and pencils, 15

*To reduce the number of disposable pipets required for a class of 30 students, dispense metal ion solutions in labeled pipet sets to be shared by several groups of students.

Preparation of Solutions *(for a class of 30 students working in pairs)*

Copper(II) Sulfate, 1 M: Dissolve 12.5 g of copper(II) sulfate pentahydrate (CuSO₄·5H₂O) in about 25 mL of distilled or deionized water. Stir to dissolve and dilute to 50 mL with water.

Iron(II) Sulfate, 1 M: Dissolve 13.9 g of iron(II) sulfate heptahydrate (FeSO₄·7H₂O) in about 25 mL of distilled or deionized water. Stir to dissolve and dilute to 50 mL with water.
Note: Prepare this solution fresh the day of lab.

Magnesium Sulfate, 1 M: Dissolve 12.3 g of magnesium sulfate heptahydrate (MgSO₄·7H₂O) in about 25 mL of distilled or deionized water. Stir to dissolve and dilute to 50 mL with water.

Silver Nitrate, 1 M: Dissolve 8.5 g of silver nitrate in about 25 mL of distilled or deionized water. Stir to dissolve and dilute to 50 mL with water.

Sodium Nitrate, 1 M: Dissolve 4.3 g of sodium nitrate in about 25 mL of distilled or deionized water. Stir to dissolve and dilute to 50 mL with water.

Zinc Sulfate, 1 M: Dissolve 14.4 g of zinc sulfate heptahydrate (ZnSO₄·7H₂O) in about 50 mL of distilled or deionized water. Stir to dissolve and dilute to 50 mL with water.

Teacher's Notes

Safety Precautions

Silver nitrate solution is a corrosive liquid and is toxic by ingestion. It will stain skin and clothing. Spray 'n Wash® laundry product may be used to remove silver nitrate stains from clothing. Copper(II) sulfate solution is toxic by ingestion. Iron(II) sulfate and zinc sulfate solutions are slightly toxic. Magnesium metal is a flammable solid; avoid contact with flames and heat. Metal pieces may have sharp edges—handle with care. Avoid contact of all chemicals with eyes and skin. Please review current Material Safety Data Sheets for additional safety, handling, and disposal information. Remind students to wash their hands thoroughly with soap and water before leaving the lab.

Disposal

Consult your current *Flinn Scientific Catalog/Reference Manual* for general guidelines and specific procedures governing the disposal of laboratory waste. Iron(II) sulfate solution is air and light sensitive—prepare the solution fresh for each lab. Excess iron(II) sulfate solution may be disposed of by rinsing it down the drain with plenty of excess water according to Flinn Suggested Disposal Method #26b. Save the other metal ion solutions in properly labeled bottles for future use. With the exception of magnesium, the metal pieces may be reused from lab to lab and from year to year. Rinse the metal pieces with distilled water, dry them thoroughly on paper towels, and store them in properly labeled zipper-lock plastic bags for future use. Magnesium metal ribbon should be freshly cut each year; dispose of used magnesium pieces in the solid trash according to Flinn Suggested Disposal Method #26a.

Lab Hints

- The laboratory work and calculations for this microscale experiment can be completed in a typical 50-minute lab period. Beginning students often forget to record which metal is connected to the positive or negative lead and then struggle with identifying the anode and cathode. Fortunately, the micro-voltaic cells are endlessly reusable and the measurements are quick and easy. Encourage students to repeat their measurements if they are unsure of their results.

- Perform the following simple demonstration to show the relationship between a spontaneous redox reaction and the identity of the cathode and the anode in a voltaic cell. Place a strip of copper metal in a solution of zinc sulfate in one beaker, and a strip of zinc metal in a solution of copper sulfate in a second beaker. Only one reaction is spontaneous—the reaction of zinc metal with Cu(II) ions. A positive cell voltage will be obtained when copper is the cathode (attached to the positive lead on the voltmeter) and zinc is the anode (attached to the negative lead on the voltmeter).

- It may save valuable lab time for the teacher to sand larger sheets or strips of metal prior to class. With the exception of magnesium, however, most of the metal pieces may be stored and reused from year to year. Students should re-polish these metal pieces each year. To avoid mixing different metals that look similar (e.g., Zn and Fe), cut these metals into different shapes.

- See the *Master Materials Guide* at the end of this book for the recommended forms of the metals and their Flinn Catalog numbers. Metal strips or sheet gave better results than metal foil.

Teacher's Notes

Teacher Notes

- Other metal ion solutions may be substituted for the metal sulfate solutions—try copper(II) nitrate, copper(II) chloride, iron(II) ammonium sulfate, magnesium nitrate, zinc nitrate, etc. Assuming the concentration of metal ions in each microcell is equal, the actual concentration (e.g., 0.1, 0.5 or 1 M) should not affect the measured cell potentials. We have found, however, that the most consistent results were obtained using 1 M solutions of the metal sulfates, as described in the *Materials* section.

- Either sodium nitrate or potassium nitrate may be used as the "salt bridge" electrolyte in this experiment. Do not use sodium or potassium chloride as the electrolyte if silver is one of the metals being tested.

- Other metals may also be used in this experiment. If time permits, consider adding nickel, tin, lead or aluminum to the list of metals to be tested. Students may set up two filter paper cutouts using a total of nine different metals (zinc must be repeated as the reference electrode on each filter paper). Use nickel sulfate, lead(II) nitrate, tin(II) chloride, and aluminum sulfate solutions as the source of metal ions for their respective half-cells. Tin and lead usually give very good results. Aluminum tends to give poor results, due to its "self-protecting" oxide coating. *Note:* Using heavy metal salt solutions may require dedicated heavy metal waste disposal.

- We tested several variations of the procedure for measuring cell potentials. Half-cells were constructed in 50- or 100-mL beakers and in microscale reaction plates, with salt bridges provided by strips of filter paper soaked in salt solution. The micro-voltaic cell procedure recommended in this write-up gave the most consistent results and was easiest to work with. Microscale reaction plates gave good results, but it was more difficult to hold the metals pieces stable in the wells when touching them with the instrument leads (especially with the voltage probe). Larger half-cells in 50-mL beakers generally gave less accurate results.

- See "Metal Activity and Reactivity" in *Oxidation and Reduction,* Volume 16 in the *Flinn ChemTopic™ Labs* series, for a complementary study of metal activity using single replacement reactions.

Teaching Tip

- Many textbooks use the equation $E°_{cell} = E°_{red}$ (cathode) $+ E°_{ox}$ (anode) to calculate cell potentials. This equation may foster the misconception that half-cell potentials are additive (like ΔH values, etc.). This is not the case. Standard reduction potentials for individual half-cell reactions have no intrinsic or absolute meaning. *Cell potentials are only defined by difference.* That is why it is better to express the cell potential as the difference between the standard reduction potential of the cathode versus the anode (see Equation 1 in the *Background* section).

Measuring Cell Potentials

Teacher's Notes

Answers to Pre-Lab Questions *(Student answers will vary.)*

Standard reduction potentials ($E°_{red}$) are defined for *reduction* half-reactions. For a voltaic cell, $E°_{red}$ for the cathode (reduction) must be more positive than $E°_{red}$ for the anode (oxidation). Consider the following two metals and their standard reduction potentials.

$$Co^{2+}(aq) + 2e^- \rightarrow Co(s) \qquad E°_{red} = -0.28 \text{ V}$$

$$Ag^+(aq) + e^- \rightarrow Ag(s) \qquad E°_{red} = 0.80 \text{ V}$$

1. If half-cells for these two metals were combined in a voltaic cell, what reaction would take place at the anode? What reaction would take place at the cathode?

    The half-cell with the more positive standard reduction potential will be the cathode. The half-cell with the less positive or negative standard reduction potential will be the anode.

    $$Ag^+(aq) + e^- \rightarrow Ag(s) \qquad \textit{Cathode—reduction}$$

    $$Co(s) \rightarrow Co^{2+}(aq) + 2e^- \qquad \textit{Anode—oxidation}$$

2. Calculate the cell potential for the overall (spontaneous) reaction under standard conditions.

    $$E°_{cell} = E°_{red}(cathode) - E°_{red}(anode) = 0.80 \text{ V} - (-0.28 \text{ V}) = 1.08 \text{ V}$$

3. Write the balanced chemical equation for the overall reaction.

    $$2Ag^+(aq) + Co(s) \rightarrow 2Ag(s) + Co^{2+}(aq)$$

4. The activity of a metal is defined as its ease of oxidation (an active metal is one that is easily oxidized). Based on their standard reduction potentials, which is the more active metal, cobalt or silver?

    Cobalt is more active than silver—it has a more negative reduction potential.

    Note to teachers: *Lithium has the most negative standard reduction potential. Other alkali metals, however, are easier to oxidize than lithium. The apparent discrepancy between the the value of $E°_{red}$ for lithium and its ease of oxidation is due to the extremely small size of the Li^+ ion.*

Teacher's Notes

Teacher Notes

Sample Data

Student data will vary.

Data Table

Part A. Cell Potentials versus Zinc as the Reference Electrode

Metal (M)	Positive Electrode (Cathode)	Negative Electrode (Anode)	Measured Cell Potential $E°_{cell}$	Calculated Value $E°_{red}$ (M)
Copper	Copper	Zinc	+1.05 V	+0.29 V (Cu)
Iron	Iron	Zinc	+0.36 V	–0.40 V (Fe)
Magnesium	Zinc	Magnesium	+0.77 V	–1.53 V (Mg)
Silver	Silver	Zinc	+1.46 V	+0.70 V (Ag)

Part B. Predicted and Measured Cell Potentials*

Cathode/Anode	Calculation (Equation 1)	Predicted Cell Potential	Measured Cell Potential
Copper/Iron	$E°_{cell} = 0.29\ V - (-0.40\ V)$	+0.69 V	+0.66 V
Copper/Magnesium	$E°_{cell} = 0.29\ V - (-1.53\ V)$	+1.82 V	+1.77 V
Silver/Copper	$E°_{cell} = 0.70\ V - 0.29\ V$	+0.41 V	+0.46 V
Iron/Magnesium	$E°_{cell} = -0.40\ V - (-1.53\ V)$	+1.13 V	+0.85 V
Silver/Iron	$E°_{cell} = 0.70 - (-0.40\ V)$	+1.10 V	+1.14 V
Silver/Magnesium	$E°_{cell} = 0.70 - (-1.53\ V)$	+2.23 V	+1.92 V

*The results for all six possible half-cell combinations are given here. Students were asked to measure two cell potentials in Part B. Class results should be posted for all six cells in Part B so that students may answer Post-Lab Question #2.

Answers to Post-Lab Questions *(Student answers will vary.)*

1. Write the balanced chemical equation for the overall reaction in each voltaic cell in Part A.

$$Zn(s) + Cu^{2+}(aq) \rightarrow Zn^{2+}(aq) + Cu(s)$$

$$Zn(s) + Fe^{2+}(aq) \rightarrow Zn^{2+}(aq) + Fe(s)$$

$$Mg(s) + Zn^{2+}(aq) \rightarrow Mg^{2+}(aq) + Zn(s)$$

$$Zn(s) + 2Ag^{+}(aq) \rightarrow Zn^{2+}(aq) + 2Ag(s)$$

Measuring Cell Potentials

Teacher's Notes

2. Consider the class results for all possible cell combinations in Parts A and B: Which metal was most easily oxidized (it always appeared as the anode)? Which metal ions were most easily reduced (the corresponding metal always appeared as the cathode)?

    Magnesium was the most easily oxidized metal tested in this experiment (magnesium always acted as the anode). Silver ions were the most easily reduced (silver always acted as the cathode).

3. Rank the five metals tested (including zinc) from most positive to most negative standard reduction potential. Write a general statement describing the relationship between the standard reduction potential of a metal and metal activity.

    From most positive to most negative reduction potential:

    Ag > Cu > Fe > Zn > Mg

    Metals with more negative standard reduction potentials are more active (more easily oxidized) than metals with more positive standard reduction potentials. Magnesium was the most active metal tested, and it had the most negative standard reduction potential.

4. Look up the literature values of the standard reduction potentials for Cu, Fe, Mg, and Ag and calculate the percent error for each. *Hint:* Note the symbol for "absolute value."

$$\text{Percent error} = \frac{|\text{Experimental value} - \text{Literature value}|}{\text{Literature value}} \times 100\%$$

Standard Reduction Potentials

Metal	Experimental Value	Literature Value	Percent Error
Cu	+0.29 V	+0.34 V	15%
Fe	−0.40 V	−0.44 V	9%
Mg	−1.53 V	−2.36 V	35%
Ag	+0.70 V	+0.80 V	13%

Note to teachers: *Cells featuring magnesium consistently showed voltages significantly less positive than their maximum calculated values. Metals closer together in the electrochemical series tended to give the most accurate results.*

Page 1 – **Quantitative Electrochemistry**

Teacher Notes

Quantitative Electrochemistry
Coulombs, Electrons, and Moles

Introduction

Electrochemistry is the branch of chemistry that deals with the interconversion of electrical and chemical energy in chemical reactions. As the name implies, electrochemistry bridges two disciplines—electricity and chemistry. What is the relationship between the quantity of electricity and the extent of a chemical reaction in an electrochemical process?

Concepts

- Electrolysis
- Electrical charge (coulombs)
- Current (amperes)
- Faraday constant (coulombs per mole)

Background

There are two basic types of electrochemical processes. A voltaic cell is a chemical device for the production of electricity—the energy released in a spontaneous chemical reaction is changed into electrical energy. In an electrolytic cell, an electric current is passed through a solution and causes a nonspontaneous chemical reaction to occur. *Electrolysis* has many industrial and commercial applications. Sodium hydroxide, chlorine, and many metals, such as aluminum, are all produced industrially by electrolysis. The amount of electricity used in electrolysis represents a significant portion of our nation's energy consumption. The production of aluminum, for example, requires an average of 7 kilowatt hours of electricity per pound of aluminum.

The principles governing the amount of product obtained in electrolysis were developed by Michael Faraday, one of the greatest scientists of the 19th century. In 1834, Michael Faraday published a paper entitled "On Electrical Decomposition," in which he defined many of the terms that are still used today to describe electrochemical cells (electrode, cathode, anode, electrolyte, etc.). He also laid out the mathematical relationship between the quantity of electricity and the amount of a substance produced in electrolysis. According to Faraday:

- The amount of a substance deposited on each electrode in an electrolytic cell is directly proportional to the quantity of electricity passed through the cell.

- The quantity of an element deposited by a given amount of electricity depends on its chemical equivalent weight (molar mass).

Faraday's work in electrochemistry led to important advances in our understanding of both electricity and chemistry. His work has been honored in many ways, including in the name of the *Faraday constant*, a fundamental physical constant corresponding to the charge in coulombs of one mole of electrons. In this experiment, the value of the Faraday constant will be determined by measuring the amount of copper obtained in an electroplating reaction.

How much electricity is needed to make an aluminum soda can? This is a good exercise to get students thinking about the quantitative aspects of electrochemistry. One pound of aluminum requires about 7 kilowatt-hours of electricity and will make about 25 beverage cans. So, each can "consumes" about 0.28 kilowatt-hours of electricity. That's the equivalent of burning a 100-watt bulb for almost 3 hours!

Quantitative Electrochemistry – Page 2

Teacher Notes

Electroplating is the process of depositing a metal on the surface of a conductor by passing electricity through a solution of metal ions. Figure 1 shows a basic diagram of an electrolytic cell for a "copper-plating" reaction from an aqueous solution of copper(II) ions. The electrodes are copper metal. Oxidation of copper metal to copper(II) ions occurs at the anode, and reduction of copper(II) ions to copper metal occurs at the cathode.

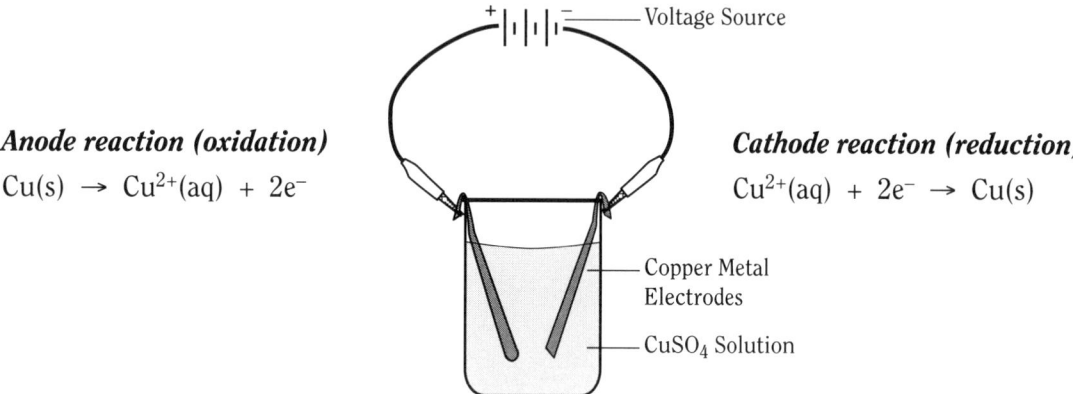

Anode reaction (oxidation)
$Cu(s) \rightarrow Cu^{2+}(aq) + 2e^-$

Cathode reaction (reduction)
$Cu^{2+}(aq) + 2e^- \rightarrow Cu(s)$

Figure 1. Basic Diagram of an Electroplating Cell.

The change in mass at each electrode depends on the reaction time and the *current*—the amount of charge that flows through the cell per second. The coulomb (C) is the fundamental unit of electrical charge, and the charge on the electron is equal to 1.602×10^{-19} C. Current is measured in amperes, where one ampere (A) is equal to the flow of one coulomb of charge per second (1 A = 1 C/sec). The flow chart in Figure 2 shows the steps involved in calculating the amount of a substance produced in electrolysis as a function of the current and of the reaction time.

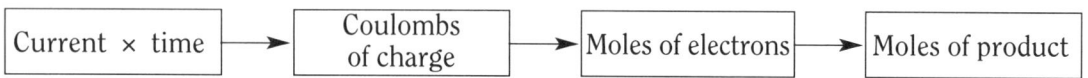

Figure 2. Flow Chart for Electrolysis Calculations.

Experiment Overview

The purpose of this experiment is to determine the average value of the Faraday constant ($\mathcal{F}$) based on Faraday's laws of electrolysis for the electroplating of copper from a copper(II) sulfate solution. The amount of copper gained or lost at each electrode will be measured, and the current and the time of electrolysis will be recorded.

Pre-Lab Questions

Calculate the amount of time it would take to plate precisely 2.00 g of copper onto the cathode in Figure 1 if the current is 1.25 A.

1. Determine the moles of copper metal and the *total number of electrons* needed to obtain this amount of Cu atoms from Cu^{2+} ions.

2. Multiply the number of electrons by the charge per electron (see the *Background* section) to find the total quantity of charge in coulombs that must be passed through the solution.

3. Divide the charge in coulombs by the current (1.25 A = 1.25 C/sec) to determine the number of seconds required. Convert the time in seconds to minutes.

Flinn ChemTopic™ Labs — Electrochemistry

Teacher Notes

Materials

Copper strips, 1 cm × 10 cm, 2
Copper(II) sulfate solution, CuSO$_4$, 1 M, 80 mL
Steel wool or sandpaper
Balance, centigram or milligram precision
Beaker, 150-mL
Distilled water and wash bottle
Paper towels
"Rinse" beaker containing distilled water*
"Rinse" beaker containing isopropyl alcohol or acetone*

Ammeter (0.2 to 1.0 A)
Alligator cords with clips, 3
Batteries, D-cell, 4, or power supply
Battery pack (to connect batteries in series)
Clock or timer
Forceps
Scissors

*Several groups of students may share alcohol and water rinse beakers.

Safety Precautions

Isopropyl alcohol is a flammable organic solvent; avoid contact with flames and heat. Copper(II) sulfate solution is toxic by ingestion and is a skin and eye irritant. Avoid contact with eyes and skin. Metal pieces may have sharp edges—handle with care. Wear chemical splash goggles, chemical-resistant gloves, and a chemical-resistant apron. Wash hands thoroughly with soap and water before leaving the lab.

Procedure

1. Connect four D-cell (1.5 V) batteries in series using a battery pack or battery holders. *Note:* The battery pack consists of a molded plastic sleeve that will hold four batteries. The pack includes metal plates that may be placed at either end to connect the batteries to the electrochemical cell.

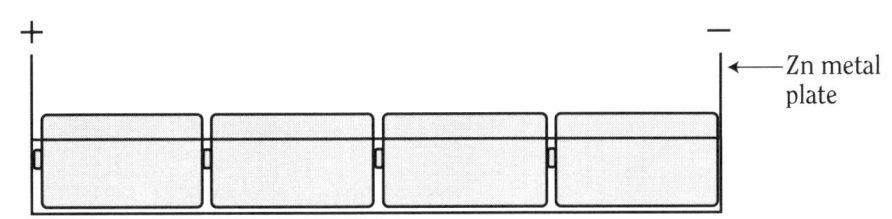

Figure 3. Battery Pack with Four D-Cells in Series.

2. Obtain two 1 cm × 10 cm strips of copper metal. Using scissors, round off the top edge of one of the copper strips. *The copper strip with the rounded edge will be used as the anode in this experiment.*

3. Polish the copper strips with steel wool or sandpaper and wipe clean with a paper towel.

4. Rinse each copper strip with a gentle stream of distilled water from a wash bottle, and pat dry.

5. Holding the metal strips with forceps, dip each copper electrode into a beaker containing isopropyl alcohol or acetone. Remove the copper strips and allow to air dry.

6. When the copper strips are dry, measure and record the mass of each electrode. *Remember that the strip with the rounded edge is the anode.*

The copper electrodes may be reused from one class to the next. If the copper strips have been recycled, instruct students to skip step 2—otherwise all of the strips will have rounded edges by the time the last class is ready to use them. See the Lab Hints *for more information.*

Quantitative Electrochemistry – Page 4

7. Obtain about 80 mL of 1 M copper(II) sulfate in a 150-mL beaker and place the two copper electrodes upright in the beaker. Be sure that the electrodes do not touch each other.

8. Set up the electroplating apparatus (Figure 4): Using alligator clips, connect the anode to the positive lead on the battery pack, and the cathode to the positive lead on the ammeter.

9. Check the apparatus, then connect the negative lead of the ammeter to the negative lead of the battery pack. *Record the initial time.*

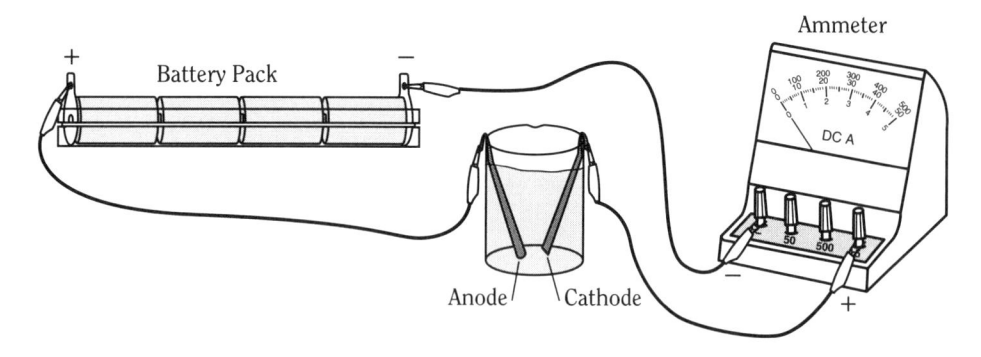

Figure 4. Electroplating Apparatus.

10. Bend the tops of the electrodes over the sides of the beaker to hold the electrodes in position. Read the current on the ammeter for a few seconds. The current will depend on the distance between the electrodes.

11. Move the electrodes closer together or farther apart as necessary to obtain a stable current in the range 0.6–0.7 A. The maximum current is usually obtained with the electrodes about 1–2 cm apart. *Do not allow the electrodes to touch.* After adjusting the electrodes, try not to move the electrodes during the electrolysis, because this will affect the current flow.

12. Measure and record the current at one minute intervals throughout the electrolysis reaction.

13. Continue electrolysis for 18–20 minutes, as time permits. *Record the final time at the end of the electrolysis reaction.*

14. Disconnect the alligator clips from the electrodes and gently remove the electrodes from the beaker.

15. Using forceps, carefully dip each electrode into a beaker of distilled water and then into a beaker of isopropyl alcohol or acetone to rinse the electrodes. Place the electrodes on a paper towel and allow to air dry. Do not wipe the surfaces of the electrodes!

16. When the copper strips are dry, measure and record the mass of each electrode. Remember that the strip with the rounded edge is the anode.

17. Consult your instructor concerning disposal of the copper(II) sulfate solution and the copper electrodes.

Teacher Notes

Steps 10 and 11 require some manual dexterity on the part of the students to prevent the electrodes from touching and to obtain the maximum (stable) current. To hold the electrodes in place, cut slits for the electrodes on a piece of acetate transparency and place the transparency on top of the beaker.

Teacher Notes

Name: _____

Class/Lab Period: _____

Quantitative Electrochemistry

Data Table

Mass of Anode (Initial)	
Mass of Cathode (Initial)	
Mass of Anode (Final)	
Mass of Cathode (Final)	
Time (Initial)	
Time (Final)	
Total Electrolysis Time	

Record current at 1 minute intervals:

1 min		11 min	
2 min		12 min	
3 min		13 min	
4 min		14 min	
5 min		15 min	
6 min		16 min	
7 min		17 min	
8 min		18 min	
9 min		19 min	
10 min		20 min	

Quantitative Electrochemistry – Page 6

Post-Lab Calculations and Analysis

(Show all work on a separate sheet of paper. Enter the results of the calculations in the Results Table.)

1. Calculate the change in mass for both the anode and the cathode. Divide the change in mass by the atomic mass of copper to determine the *moles of copper* lost or gained at the anode and cathode, respectively.

2. Multiply the number of moles of copper by two to find the *moles of electrons* transferred at each electrode during electrolysis.

3. Determine the average current during the electrolysis procedure. Multiply the average current by the time of electrolysis in seconds to calculate the *coulombs of charge* passed through the cell.

4. Divide the total coulombs of charge by the charge on an electron (1.60×10^{-19} C/electron) to determine the *number of electrons* passed through the cell during electrolysis.

5. Divide the coulombs of charge passed through the cell (Question #3) by the moles of electrons transferred to each electrode to calculate the *Faraday constant, $\mathcal{F}$* (coulombs of charge per mole of electrons).

6. Look up the literature value of the Faraday constant and calculate the *percent error* in the experimental value for both the anode and the cathode. Which electrode gave more accurate results? Suggest a possible reason for any difference in accuracy.

Results Table

	Anode	Cathode
Change in Mass		
Moles of Copper		
Moles of Electrons		
Average Current*		
Coulombs of Charge*		
Number of Electrons*		
Faraday Constant		
Percent Error (Faraday Constant)		

*These values will be the same for both the anode and the cathode.

Teacher Notes

Teacher's Notes
Quantitative Electrochemistry

Master Materials List *(for a class of 30 students working in pairs)*

Copper strips, 1 cm × 10 cm, 30
Copper(II) sulfate solution, $CuSO_4$, 1 M, 1.5 L
Isopropyl alcohol or acetone, 600 mL*
Steel wool or sandpaper
Balances, centigram or milligram precision, 3
Beakers, 150-mL, 15; and 400-mL, 6*
Distilled water and wash bottles, 15
Paper towels

Ammeters (0.2 to 1.0 A), 15
Alligator cords with clips, 45
Batteries, size D, 60, or power supply, 15
Battery packs (to connect batteries in series), 15
Clock or timers, 15
Forceps, 15
Scissors

*Place several 400-mL rinse beakers containing 200 mL isopropyl alcohol or acetone in convenient locations for students to share. Water rinse beakers are also needed.

Preparation of Solution *(for a class of 30 students working in pairs)*

Copper(II) Sulfate, 1 M: Dissolve 250 g of copper(II) sulfate pentahydrate ($CuSO_4 \cdot 5H_2O$) in about 500 mL of distilled or deionized water. Stir to dissolve and dilute to 1 L with water. Repeat as needed.

Safety Precautions

Isopropyl alcohol is a flammable organic solvent; avoid contact with flames and heat. Copper(II) sulfate solution is toxic by ingestion and is a skin and eye irritant. Avoid contact with eyes and skin. Metal pieces may have sharp edges—handle with care. Wear chemical splash goggles, chemical-resistant gloves, and a chemical-resistant apron. Please review current Material Safety Data Sheets for additional safety, handling, and disposal information. Remind students to wash their hands thoroughly with soap and water before leaving the lab.

Disposal

Consult your current *Flinn Scientific Catalog/Reference Manual* for general guidelines and specific procedures governing the disposal of laboratory waste. Excess copper(II) sulfate solution may be disposed of by rinsing it down the drain with plenty of excess water according to Flinn Suggested Disposal Method #26b. Save the copper metal strips for future use. The copper strips may be cleaned by dipping them quickly in 6 M nitric acid. Rinse thoroughly with distilled water and with alcohol. Allow to air dry.

Lab Hints

- The laboratory work for this experiment can be completed in a typical 50-minute lab period. Set up and demonstrate a model apparatus before class to help students prepare for lab. This experiment is not intended as an introductory level experiment in electrochemistry. Students should have a good working knowledge of the language of electrochemistry (electrode, cathode, anode, etc.) and the signs of the electrodes. Students should also be familiar with the definitions and units used in electrical measurements (volts, current, charge, etc.).

In general, the batteries can only be used by two classes before they have been drained of power and must be replaced. A large, 6-V lantern battery may be used instead of 4 D-cells. The 6-V batteries cost more, but they will also last longer.

Teacher's Notes

- Both the copper electrodes and the copper sulfate solution may be recycled and reused by several lab sections. In our testing, we used the copper electrodes about 3–4 times without any problems—sand and polish well before each use. (The anode may become too thin after repeated use.) It may, in fact, be beneficial to reuse the electrodes from class to class. Bending and cutting a metal produces defects or stress areas in the crystal structure. Stress areas in the metal will create potential differences and may cause uneven plating or the growth of dendritic crystals that do not adhere well to the electrode. Repeated use removes these stress areas. Dispose of the copper sulfate solution if you see crystals of copper floating in the solution.

- The amount of copper gained or lost at each electrode depends on the current and the time of electrolysis. The current, in turn, depends on the voltage and the resistance of the cell, which is difficult to control. DC power supplies, if available, give better results than batteries, because the voltage can be adjusted to give a higher and more stable current. (Your colleagues in the physics department may have power supplies that you can borrow.) The highest, stable voltage we could achieve using four D-cell batteries in series (6 V) was about 0.7 A. At this current level, an electrolysis time of 20 minutes is recommended to give accurate results with a centigram balance. If power supplies and/or milligram balances are available, the electrolysis time may be reduced to 15 minutes without sacrificing precision. (Otherwise, the uncertainty in the balance measurement will be a significant percentage of the change in mass.) A maximum current of about 1 A is recommended if a power supply will be used.

- The electrolytic cell and the ammeter are set up in series. The total resistance of the circuit is equal to the sum of the resistances supplied by each component. Using a variable resistor (rheostat) should, in theory, give a more stable current and improve the results. We have found, however, that using a rheostat introduces another variable and does not give better accuracy. The rheostat tends to heat up when it is connected to the voltage source. In order to avoid current fluctuations, the rheostat must be connected to the power supply and allowed to heat up for about 5 minutes before use.

- Some sources recommend adding sulfuric acid to the electrolysis solution (making it 1 M in $CuSO_4$ and 1 M in H_2SO_4) to decrease the resistance of the cell. We tested this modification and found that there was no difference in accuracy with and without sulfuric acid.

Teaching Tip

- The first "instrument" to measure the amount of electrical current flowing through a circuit was the volta-electrometer, which was designed by Michael Faraday for his quantitative studies of electrolysis. The volta-electrometer consisted of an electrolytic cell for the decomposition of water—the oxygen and hydrogen generated in the reaction were collected and measured. Faraday used the quantity of electricity that would generate 1 g of hydrogen gas as the reference standard. He found that the amount of a substance generated by this quantity of electricity was equal to its "electrochemical equivalent" (eight for oxygen, 36 for chlorine, 104 for lead, etc.). In modern terms, the electrochemical equivalent is the molar mass of the element divided by the number of electrons gained or lost in electrolysis. Faraday's laws of electrolysis appear even more remarkable when we remember that he worked all of this out more than 50 years before the electron was discovered!

Teacher's Notes

Teacher Notes

Answers to Pre-Lab Questions *(Student answers will vary.)*

Calculate the amount of time it would take to plate precisely 2.00 g of copper onto the cathode in Figure 1 if the current is 1.25 A.

1. Determine the moles of copper metal and the *total number of electrons* needed to obtain this amount of Cu atoms from Cu^{2+} ions.

$$2.00 \text{ g Cu} \times \frac{1 \text{ mole Cu}}{63.55 \text{ g}} = 0.0315 \text{ moles Cu}$$

The mole ratio requires 2 moles of electrons per mole of copper metal obtained.

$$0.0315 \text{ moles Cu} \times \frac{2 \text{ moles e}^-}{\text{mole Cu}} = 0.0630 \text{ moles electrons}$$

$$0.0630 \text{ moles electrons} \times (6.02 \times 10^{23} \text{ electrons/mole}) = 3.79 \times 10^{22} \text{ electrons}$$

2. Multiply the number of electrons by the charge per electron (see the *Background* section) to find the total quantity of charge in coulombs that must be passed through the solution.

$$3.79 \times 10^{22} \text{ electrons} \times (1.602 \times 10^{-19} \text{ C/electron}) = 6.07 \times 10^{3} \text{ C}$$

3. Divide the charge in coulombs by the current (1.25 A = 1.25 C/sec) to determine the number of seconds required. Convert the time in seconds to minutes.

$$\frac{6.07 \times 10^{3} \text{ C}}{1.25 \text{ C/sec}} = 4.86 \times 10^{3} \text{ sec}$$

$$4.86 \times 10^{3} \text{ sec} \times \frac{1 \text{ min}}{60 \text{ sec}} = 81 \text{ min (rounded to the nearest minute)}$$

Teacher's Notes

Sample Data

Student data will vary.

Data Table

Mass of Anode (Initial)	2.62 g
Mass of Cathode (Initial)	3.03 g
Mass of Anode (Final)	2.36 g
Mass of Cathode (Final)	3.30 g
Time (Initial)	9:05
Time (Final)	9:25
Total Electrolysis Time	20 min

Record current at 1 minute intervals:

1 min	1.00 A	11 min	0.65 A
2 min	0.45 A	12 min	0.65 A
3 min	0.69 A	13 min	0.63 A
4 min	0.68 A	14 min	0.63 A
5 min	0.68 A	15 min	0.65 A
6 min	0.65 A	16 min	0.65 A
7 min	0.60 A	17 min	0.65 A
8 min	0.65 A	18 min	0.63 A
9 min	0.63 A	19 min	0.63 A
10 min	0.63 A	20 min	0.65 A

Answers to Post-Lab Calculations and Analysis *(Student answers will vary.)*

1. Calculate the change in mass for both the anode and the cathode. Divide the change in mass by the atomic mass of copper to determine the *moles of copper* lost or gained at the anode and cathode, respectively.

    Mass gain at cathode = 3.30 g − 3.03 g = 0.27 g

    Mass loss at anode = 2.36 g − 2.62 g = −0.26 g

    Moles of copper gained at cathode = 0.27 g/(63.55g/mole) = 0.0042 moles Cu

    Moles of copper lost at anode = 0.26 g/(63.55 g/mole) = 0.0041 moles Cu

Teacher's Notes

Teacher Notes

2. Multiply the number of moles of copper by two to find the *moles of electrons* transferred at each electrode during electrolysis.

$$\text{Moles of electrons at anode} = 0.0041 \text{ moles Cu} \times \frac{2 \text{ moles of } e^-}{\text{mole of Cu}} = 0.0082 \text{ moles } e^-$$

$$\text{Moles of electrons at cathode} = 0.0042 \text{ moles Cu} \times \frac{2 \text{ moles of } e^-}{\text{mole of Cu}} = 0.0084 \text{ moles } e^-$$

3. Determine the average current during the electrolysis procedure. Multiply the average current by the time of electrolysis in seconds to calculate the *coulombs of charge* passed through the cell.

$$\text{Average current} = 0.65 \text{ A } (0.65 \text{ C/sec})$$

$$(0.65 \text{ C/sec}) \times 1200 \text{ sec} = 780 \text{ C}$$

4. Divide the total coulombs of charge by the charge on an electron (1.60×10^{-19} C/electron) to determine the *number of electrons* passed through the cell during electrolysis.

$$\frac{780 \text{ C}}{1.60 \times 10^{-19} \text{ C/electron}} = 4.9 \times 10^{21} \text{ electrons}$$

5. Divide the coulombs of charge passed through the cell (Question #3) by the moles of electrons transferred to each electrode to calculate the *Faraday constant, $\mathcal{F}$* (coulombs of charge per mole of electrons).

$$\text{At the anode: } \frac{780 \text{ C}}{0.0082 \text{ moles}} = 9.5 \times 10^4 \text{ C/mole}$$

$$\text{At the cathode: } \frac{780 \text{ C}}{0.0084 \text{ moles}} = 9.3 \times 10^4 \text{ C/mole}$$

6. Look up the literature value of the Faraday constant and calculate the *percent error* in the experimental value for both the anode and the cathode. Which electrode gave more accurate results? Suggest a possible reason for any difference in accuracy.

$$\text{Percent error} = \frac{|\text{Literature value} - \text{Experimental value}|}{\text{Literature value}} \times 100$$

See the Sample Results Table for the results of the calculations.

Faraday constant = 9.65×10^4 C/moles.

Both the anode and the cathode gave excellent results (1–4% error) for the value of the Faraday. If the copper formed at the cathode does not adhere well to the metal surface, the mass gain at the cathode will be less accurate than the mass loss at the anode. (That's why it is important not to wipe the surfaces of the electrodes—see step 15 in the Procedure.*) Competing reduction of water during the electroplating reaction may give rise to hydrogen gas at the cathode. This would make the anode results more accurate. (Oxygen generation at the anode is less likely.)*

Teacher's Notes

Sample Results Table
Student results will vary.

	Anode	Cathode
Change in Mass	0.26 g	0.27 g
Moles of Copper	0.0041 moles	0.0042 moles
Moles of Electrons	0.0082 moles	0.0084 moles
Average Current	\multicolumn{2}{c}{0.65 A}	
Coulombs of Charge	780 C	
Number of Electrons	4.9×10^{21}	
Faraday Constant	9.5×10^4 C/mole	9.3×10^4 C/mole
Percent Error (Faraday Constant)	1.6%	3.6%

Teacher Notes

Electrolysis Reactions
Oxidation and Reduction

Introduction

Electrolysis is defined as the decomposition of a substance by means of an electric current. When an electric current is passed through an aqueous solution of sodium sulfate, the water molecules decompose via an oxidation–reduction reaction. Oxygen gas is generated at the anode, hydrogen gas at the cathode. The sodium sulfate acts as an electrolyte, increasing the current flow through the solution. Depending on the nature of the electrolyte, different reactions may take place at the anode and the cathode during the electrolysis of an aqueous solution.

Concepts

- Electrolysis
- Anode and cathode
- Oxidation and reduction
- Cell potential

Background

An electrolytic cell consists of a source of direct electrical current connected to two electrodes in a solution of an electrolyte or in a molten salt solution. The electrodes act as external conductors and provide surfaces at which electron transfer will take place. Electrons flow from the anode, which is the site of oxidation, to the cathode, which is the site of reduction. The battery or other voltage source serves as an electron "pump," pushing electrons into the electrode from the negative pole and pulling electrons from the electrochemical cell at the positive pole. The negative electrode, where the electrons enter the electrolysis setup, is the cathode. The electrons are "consumed" in a reduction half-reaction. Electrons are generated at the anode, the positive electrode, via an oxidation half-reaction. The migration of ions in the electrolyte solution completes the electrical circuit.

The following half-reactions occur in the electrolysis of water.

Oxidation half-reaction (anode) $\quad 2H_2O(l) \rightarrow O_2(g) + 4H^+(aq) + 4e^-$

Reduction-half-reaction (cathode) $\quad 2H_2O(l) + 2e^- \rightarrow H_2(g) + 2OH^-(aq)$

Electrolysis of an aqueous salt solution may generate products other than oxygen or hydrogen if the salt contains ions that are more easily oxidized or reduced than water molecules. The electrolysis of aqueous silver nitrate ($AgNO_3$), for example, generates oxygen at the anode and silver metal at the cathode. The products of the reaction demonstrate that reduction of silver ions (Ag^+) to silver (Ag) occurs more readily than reduction of water. The overall reaction is the sum of the oxidation and reduction half-reactions.

Oxidation half-reaction (anode) $\quad 2H_2O(l) \rightarrow O_2(g) + 4H^+(aq) + 4e^-$

Reduction-half-reaction (cathode) $\quad 4Ag^+(aq) + 4e^- \rightarrow 4Ag(s)$

Overall reaction $\quad 2H_2O(l) + 4Ag^+ \rightarrow O_2(g) + 4Ag(s) + 4H^+(aq)$

Electrolysis Reactions – Page 2

Experiment Overview

The purpose of this experiment is to determine the products obtained in the electrolysis of aqueous potassium iodide, copper(II) bromide, and sodium chloride solutions. The electrolysis of salt solutions will be investigated using a "Petri dish electrolysis" setup with a 9-V battery and carbon (pencil lead) electrodes (Figure 1).

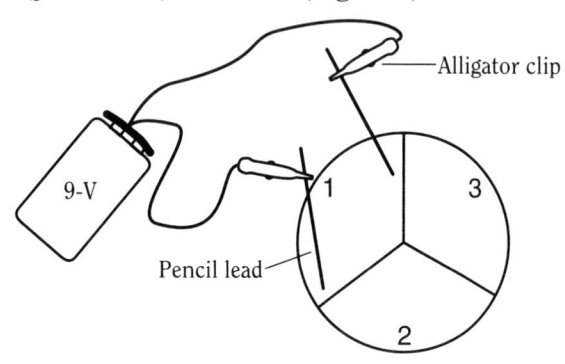

Figure 1. Petri Dish Electrolysis.

Pre-Lab Questions

1. Complete the following table summarizing the general properties of the electrodes in an *electrolytic cell*.

Electrode	Oxidation or Reduction	Sign of Electrode
Anode		
Cathode		

2. Sodium metal is produced commercially by the electrolysis of molten sodium chloride. The by-product of the reaction is chlorine gas. (a) Write the oxidation and reduction half-reactions for the electrolysis of molten sodium chloride. (b) Identify the substance that is oxidized and the substance that is reduced. (c) Write the balanced chemical equation for the overall reaction.

3. Sodium metal is easily oxidized—it is a very reactive metal. Sodium reacts spontaneously with water at room temperature to give sodium hydroxide and hydrogen gas. Would you expect to observe sodium metal in the electrolysis of aqueous sodium chloride? Explain.

Materials

Copper(II) bromide solution, $CuBr_2$, 0.2 M, 8 mL
Phenolphthalein indicator solution, 0.5%, 1 mL
Potassium iodide solution, KI, 0.5 M, 8 mL
Sodium chloride solution, NaCl, 0.5 M, 8 mL
Sodium thiosulfate "waste beaker," $Na_2S_2O_3$, 3 M in H_2SO_4 (for disposal)
Starch solution, 0.5%, 1 mL
Distilled water and wash bottle

Battery, 9-V
Battery cap w/ alligator clip leads
Beral-type pipets, 3
Paper towels
Pencil lead electrodes, 0.9-mm, 2
Petri dish, partitioned, 3-way
Stirring rod
Wax pencil or marking pen

Teacher Notes

The carbon "pencil lead" electrodes are needed for the electrolysis of aqueous potassium iodide and copper(II) bromide. The copper alligator clips may be used directly as the electrodes for the electrolysis of sodium chloride solution, but they will quickly corrode.

Teacher Notes

Safety Precautions

Copper(II) bromide solution is toxic by ingestion and may be irritating to the eyes, skin, and respiratory tract. Phenolphthalein is an alcohol-based solution and is a flammable liquid. Keep away from flames and heat. The electrolysis reactions will generate small amounts of hazardous gases. Do not breathe the vapors. Avoid contact of all chemicals with eyes and skin. Wear chemical splash goggles, chemical-resistant gloves, and a chemical-resistant apron. Wash hands thoroughly with soap and water before leaving the lab.

Procedure

1. Place the partitioned Petri dish on a sheet of white paper. Observe that the compartments or segments of the Petri dish are labeled 1, 2, and 3. *Note:* If the compartments are not labeled, label the paper underneath the Petri dish.

2. Carefully pour about 8 mL of 0.5 M potassium iodide solution into the first compartment of the Petri dish until the bottom of the compartment is covered with solution. The dish should be one-third to one-half full.

3. Add 3 drops of phenolphthalein solution and stir to mix.

4. Connect the battery cap to the 9-V battery. Carefully attach a "pencil lead" electrode to each alligator clip lead. *Caution:* Do not allow the electrodes to touch each other.

5. Hold the red (+) lead from the 9-V battery in one hand and the black (−) lead in the other hand. Keeping the electrodes as far apart as possible, dip the pencil lead electrodes into the potassium iodide solution in the Petri dish.

6. Let the electric current run for 1–2 minutes while observing any changes in the potassium iodide solution.

7. Record all observations in the data table—be sure to indicate where changes take place (at the anode or the cathode). Refer to the *Background* section and the *Pre-Lab Questions* for the properties of the electrodes.

8. Remove the pencil lead electrodes from the electrolysis solution. Carefully rinse the electrodes with distilled water from a wash bottle and gently pat dry on a paper towel.

9. Add two drops of starch solution to the potassium iodide solution after electrolysis and record observations in the data table.

10. Carefully pour about 8 mL of 0.5 M sodium chloride solution into the second compartment of the Petri dish. Add three drops of phenolphthalein indicator solution and stir to mix.

11. Repeat steps 5–8 for the electrolysis of sodium chloride solution. Record observations in the data table.

Electrolysis Reactions – Page 4

Teacher Notes

12. After electrolysis, add 3 drops of potassium iodide solution, followed by one drop of starch, to the sodium chloride solution. Record observations in the data table.

13. Carefully pour about 8 mL of 0.2 M copper(II) bromide solution into the third compartment of the Petri dish.

14. Repeat steps 5–8 for the electrolysis of copper(II) bromide solution. Record observations in the data table.

15. Remove the pencil lead electrodes from the alligator clips and disconnect the battery cap from the battery.

16. The electrolysis products may include dilute halogen solutions (chlorine, bromine, and iodine). *Working in the hood,* carefully pour the contents of the Petri dish into a waste beaker containing sodium thiosulfate solution. Sodium thiosulfate will reduce the halogen waste products. Allow the beaker to stand in the hood overnight.

Teacher Notes

Name: _____

Class/Lab Period: _____

Electrolysis Reactions

Data Table

Electrolyte (Salt Solution)	Observations	
	Anode	Cathode
Potassium Iodide		
Sodium Chloride		
Copper(II) Bromide		

Post-Lab Questions

1. The following oxidation and reduction half-reactions are possible for the electrolysis of potassium iodide solution. The solution contains water molecules, potassium ions (K$^+$), and iodide ions (I$^-$).

   $2H_2O(l) \rightarrow O_2(g) + 4H^+(aq) + 4e^-$ $2H_2O(l) + 2e^- \rightarrow H_2(g) + 2OH^-(aq)$

   $K^+(aq) + e^- \rightarrow K(s)$ $2I^-(aq) \rightarrow I_2(s) + 2e^-$

   (a) What product was formed at the anode in the electrolysis of potassium iodide solution? Explain, citing specific evidence from your observations.

   (b) What product was formed at the cathode in the electrolysis of potassium iodide solution? Explain based on your observations.

   (c) Write the balanced chemical equation for the overall redox reaction in the electrolysis of aqueous potassium iodide. *Hint:* Remember to balance the electrons!

Electrolysis Reactions

Electrolysis Reactions – Page 6

2. *Using Question #1 as a guide:* (a) Identify the products that were formed at the anode and the cathode in the electrolysis of sodium chloride solution, giving the specific evidence for their formation. (b) Write the balanced chemical equation for the overall redox reaction.

3. *Using Question #1 as a guide:* (a) Identify the products that were formed at the anode and the cathode in the electrolysis of copper(II) bromide solution, giving the specific evidence for their formation. (b) Write the balanced chemical equation for the overall redox reaction.

4. Compare the product formed at the cathode in the electrolysis of copper(II) bromide solution versus that obtained in the electrolysis of aqueous potassium iodide or sodium chloride. Explain, based on the reactivity of the metals.

5. *(Optional)* Consult a table of standard reduction potentials ($E°_{red}$): Determine the minimum voltage necessary for the electrolysis of aqueous potassium iodide.

    Hint: $E°_{cell} = E°_{red}$ (cathode) $- E°_{red}$ (anode)

Teacher's Notes
Electrolysis Reactions

Master Materials List *(for a class of 30 students working in pairs)*

Copper(II) bromide solution, $CuBr_2$, 0.2 M, 150 mL
Phenolphthalein indicator solution, 0.5%, 15 mL
Potassium iodide solution, KI, 0.5 M, 150 mL
Sodium chloride solution, NaCl, 0.5 M, 150 mL
Sodium thiosulfate "waste beaker," $Na_2S_2O_3$,
  3 M, 250 mL in 1-L beaker*
Starch solution, 0.5%, 15 mL
Distilled water and wash bottles, 15

Batteries, 9-V, 15
Battery cap, alligator clip leads, 15
Beral-type pipets, 45
Paper towels
Pencil lead electrodes, 0.9-mm, 30
Petri dishes, partitioned, 15
Stirring rods, 15
Wax pencils or marking pens, 15

Place a sodium thiosulfate solution in the hood for disposal of the halogen–water electrolysis waste solutions. This waste beaker may be used continuously by several class sections during the day.

Preparation of Solutions *(for a class of 30 students working in pairs)*

Copper(II) Bromide, 0.2 M: Dissolve 8.9 g of copper(II) bromide ($CuBr_2$) in about 100 mL of distilled or deionized water. Stir to dissolve and dilute to 200 mL with water.

Potassium Iodide, 0.5 M: Dissolve 16.6 g of potassium iodide in about 100 mL of distilled or deionized water. Stir to dissolve and dilute to 200 mL with water.

Sodium Chloride, 0.5 M: Dissolve 5.8 g of sodium chloride in about 100 mL of distilled or deionized water. Stir to dissolve and dilute to 200 mL with water.

Sodium Thiosulfate, Acidified, 3 M: Dissolve 372 g of lab grade sodium thiosulfate pentahydrate ($Na_2S_2O_3 \cdot 5H_2O$) in about 250 mL of distilled or deionized water. Stir to dissolve and dilute to 500 mL with 1 M sulfuric acid.

Safety Precautions

Copper(II) bromide solution is toxic by ingestion and may be irritating to the eyes, skin, and the respiratory tract. Phenolphthalein is an alcohol-based solution and is a flammable liquid. Keep away from flames and heat. Sodium thiosulfate acidified solution is a body tissue irritant. The electrolysis reactions will generate small amounts of hazardous gases. Perform this experiment in a well-ventilated lab only and do not breathe the vapors. Avoid contact of all chemicals with eyes and skin. Wear chemical splash goggles, chemical-resistant gloves, and a chemical-resistant apron. Please review current Material Safety Data Sheets for additional safety, handling, and disposal information. Remind students to wash their hands thoroughly with soap and water before leaving the lab.

Teacher's Notes

Disposal

Consult your current *Flinn Scientific Catalog/Reference Manual* for general guidelines and specific procedures governing the disposal of laboratory waste. Electrolysis of aqueous potassium iodide, sodium chloride, and copper(II) bromide generates halogen–water solutions. The contents of the Petri dishes should be collected in a central waste disposal beaker located in the hood and then reduced with sodium thiosulfate according to Flinn Suggested Disposal Method #12a. The resulting waste solution should be allowed to sit overnight to thoroughly degas. It may then be rinsed down the drain with plenty of excess water according to Flinn Suggested Disposal Method #26b. Do not dispose of the electrolysis waste solutions directly down the drain.

Lab Hints

- The laboratory work for this experiment can easily be completed in a typical 50-minute lab period. The experiment works best as a follow-up to the electrolysis of water, performed as either an experiment (see "Introduction to Electrochemistry") or a demonstration (see "Hoffman Electrolysis"). Students may need help correlating the color change at the cathode in the electrolysis of potassium iodide and sodium chloride with the production of OH⁻ ions from the reduction of water molecules.

- The small amount of chlorine generated in the electrolysis of sodium chloride is noticeable only by a faint odor. The concentration is not strong enough to color the solution. The test for chlorine (step 12 in the *Procedure*) involves adding potassium iodide and starch to observe the formation of the familiar iodine–starch complex. Chlorine is a stronger oxidizing agent than iodine and therefore oxidizes iodide anions to iodine. See the experiment "All in the Family" in *The Periodic Table,* Volume 4 in the *Flinn ChemTopic™ Labs* series, for a study of the reactivity and single replacement reactions of the halogens.

- Potassium iodide solution is light- and air-sensitive. Prepare the solution fresh within two weeks of its anticipated use and store the solution in a dark bottle, if possible.

- Electrolysis of salt solutions takes place at concentrations greater than about 0.2 M. We recommend using 0.5 M solutions for potassium iodide and sodium chloride—electrolysis is very rapid and the color changes are more obvious than in 0.2 M solutions. The iodine or chlorine odor generated in these reactions is very faint. In the electrolysis of copper(II) bromide, however, the odor due to bromine is more pronounced when the salt concentration is 0.5 M. We recommend a concentration of 0.2 M for the electrolysis of copper(II) bromide. The halogen odors are not a hazard when the experiment is performed as written in a well-ventilated lab. Remind students, however, never to "sniff" their experiments!

- Other electrolytes, such as silver nitrate and zinc bromide, may also be used in this experiment. Both of these metal ions are more easily reduced than the hydrogen atoms in water. Electrolysis of silver nitrate generates silver metal at the cathode. Zinc bromide gives zinc metal at the cathode and bromine at the anode.

Teacher's Notes

Teacher Notes

- This experiment may be "supersized" by carrying out the reactions in U-tubes with carbon rod electrodes and a 6-V lantern battery as the power source. About 30–50 mL of electrolyte solution will be needed, depending on the size of the U-tubes.

- Universal indicator may be used as the acid–base indicator in the electrolysis of potassium iodide or sodium chloride.

- The *Supplementary Information* section contains instructions for building a combined battery–electrode assembly using two #2 pencils and a 9-V battery with snaps.

Teaching Tips

- If your lab schedule does not allow you to have students perform this activity as an experiment, do it as a demonstration instead. No procedural changes are necessary—simply place the partitioned Petri dish on an overhead projector and follow the instructions for a great critical-thinking exercise. Write down all of the possible oxidation and reduction half-reactions for each salt on the board, and then have students identify the actual products based on their observations. In fact, this approach may be helpful even after students have completed the lab activity. This topic is very difficult for students to understand.

- Based on standard reduction potential values, oxidation of chloride ion to chlorine ($E°_{red} = -1.36$ V) is less favorable than oxidation of water to oxygen ($E°_{red} = -1.23$ V). However, there is a significant overvoltage for the oxidation of water, and thus chlorine is observed in the electrolysis of aqueous sodium chloride solution. The cause of the overvoltage is usually ascribed to a kinetically slow reaction at the anode. $E°_{red}$ values predict the thermodynamic tendency of a reaction to occur, not how fast or slow the reaction will be.

- Have individual student groups research and then present a class seminar on (a) the historical role of electrolysis in the discovery of potassium, sodium, magnesium, calcium, strontium, and barium; or (b) the modern importance of electrolysis in the production of industrial chemicals, including aluminum, sodium hydroxide, chlorine, etc.

Teacher's Notes

Answers to Pre-Lab Questions *(Student answers will vary.)*

Teacher Notes

1. Complete the following table summarizing the general properties of the electrodes in an *electrolytic cell*.

Electrode	Oxidation or Reduction	Sign of Electrode
Anode	Oxidation	Positive
Cathode	Reduction	Negative

2. Sodium metal is produced commercially by the electrolysis of molten sodium chloride. The byproduct of the reaction is chlorine gas. (a) Write the oxidation and reduction half-reactions for the electrolysis of molten sodium chloride. (b) Identify the substance that is oxidized and the substance that is reduced. (c) Write the balanced chemical equation for the overall reaction.

   (a) *Oxidation half-reaction (anode)* $2Cl^-(l) \rightarrow Cl_2(g) + 2e^-$

   Reduction half-reaction (cathode) $Na^+(l) + e^- \rightarrow Na(l)$

   Note to teachers: Sodium metal is a liquid at the temperture required for the electrolysis of molten sodium chloride.

   (b) *Chloride anions are oxidized to chlorine gas; sodium cations are reduced to sodium metal.*

   (c) *Overall balanced equation*

   $2NaCl(l) \rightarrow 2Na(l) + Cl_2(g)$

   Note to teachers: *Remind students about the need to balance electrons as well as atoms and charge when balancing the chemical equation for a redox reaction.*

3. Sodium metal is easily oxidized—it is a very reactive metal. Sodium reacts spontaneously with water at room temperature to give sodium hydroxide and hydrogen gas. Would you expect to observe sodium metal in the electrolysis of aqueous sodium chloride? Explain.

   The fact that sodium is very reactive and easily oxidized suggests that it should be extremely difficult to reduce sodium cations. In aqueous sodium chloride solution, therefore, reduction of water to hydrogen gas should be more favorable than reduction of sodium cations to sodium metal. Sodium metal will not be generated in the electrolysis of aqueous sodium chloride.

Teacher's Notes

Teacher Notes

Sample Data

Student data will vary.

Data Table

Electrolyte (Salt Solution)	Observations — Anode	Observations — Cathode
Potassium Iodide	Yellow substance formed at positive electrode and dissolved in solution. Brownish-yellow solid observed on electrode. Solution turned black when starch was added.	Rapid gas bubbling observed at negative electrode. Solution immediately surrounding the cathode turned bright pink.
Sodium Chloride	Slow bubbling at positive electrode—very faint odor of chlorine (swimming pool smell). Solution turned dark blue when potassium iodide and starch were added.	Rapid gas bubbling observed at negative electrode. Solution immediately surrounding the cathode turned bright pink.
Copper(II) Bromide	Rapid bubbling observed at positive electrode. Solution around anode turned yellow (original color was blue-green). Strong odor.	Dark solid deposited on negative electrode. The color of the solid was not obvious until the electrode was removed from the solution—reddish-brown solid. Solid rubbed off on paper towel.

Answers to Post-Lab Questions *(Student answers will vary.)*

1. The following oxidation and reduction half-reactions are possible for the electrolysis of potassium iodide solution. The solution contains are water molecules, potassium ions (K^+), and iodide ions (I^-).

   $2H_2O(l) \rightarrow O_2(g) + 4H^+(aq) + 4e^-$ $2H_2O(l) + 2e^- \rightarrow H_2(g) + 2OH^-(aq)$

   $K^+(aq) + e^- \rightarrow K(s)$ $2I^-(aq) \rightarrow I_2(s) + 2e^-$

   (a) What product was formed at the anode in the electrolysis of potassium iodide solution? Explain, citing specific evidence from your observations.

   *The substance formed at the anode is an oxidation product. The product is yellow, water-soluble, and turns black when starch is added—**iodine**.*

   (b) What product was formed at the cathode in the electrolysis of potassium iodide solution? Explain based on your observations.

   *The substance formed at the cathode is a reduction product. The product is a gas, and is accompanied by the formation of a base (phenolphthalein turned pink). The product is **hydrogen**, and **hydroxide ions** are formed as a by-product.*

Electrolysis Reactions

Teacher's Notes

    (c) Write the balanced chemical equation for the overall redox reaction in the electrolysis of aqueous potassium iodide. *Hint:* Remember to balance the electrons!

$$2H_2O(l) + 2I^-(aq) \rightarrow H_2(g) + I_2(aq) + 2OH^-(aq)$$

2. *Using Question #1 as a guide:* (a) Identify the products that were formed at the anode and the cathode in the electrolysis of sodium chloride solution, giving the specific evidence for their formation. (b) Write the balanced chemical equation for the overall redox reaction.

    *(a) The substance formed at the anode (oxidation product) is a water-soluble gas with a "swimming pool" odor—**chlorine**. The dark yellow color observed when potassium iodide was added is due to iodine. (Chlorine oxidizes iodide ions to iodine.)*

    *The substance formed at the cathode (reduction product) is **hydrogen**. Hydroxide ions are formed as a by-product. See the observations and explanation for electrolysis of potassium iodide.*

    (b) $2H_2O(l) + 2Cl^-(aq) \rightarrow H_2(g) + Cl_2(aq) + 2OH^-(aq)$

3. *Using Question #1 as a guide:* (a) Identify the products that were formed at the anode and the cathode in the electrolysis of copper(II) bromide solution, giving the specific evidence for their formation. (b) Write the balanced chemical equation for the overall redox reaction.

    *(a) The substance formed at the anode (oxidation product) is a dark yellow, water-soluble liquid with a sharp odor—**bromine**.*

    *The substance formed at the cathode (reduction product) is a reddish brown solid— **copper metal**.*

    (b) $Cu^{2+}(aq) + 2Br^-(aq) \rightarrow Cu(s) + Br_2(aq)$

4. Compare the product formed at the cathode in the electrolysis of copper(II) bromide solution versus that obtained in the electrolysis of aqueous potassium iodide or sodium chloride. Explain, based on the reactivity of the metals.

    Copper metal was obtained at the cathode (reduction product) in the electrolysis of copper(II) bromide solution. This contrasts with the formation of hydrogen as the reduction product in the electrolysis of aqueous potassium iodide or sodium chloride. Copper(II) ions are therefore more easily reduced than water molecules or potassium or sodium ions (Ease of reduction: $Cu^{2+} > H_2O \gg Na^+, K^+$). Potassium and sodium are very reactive metals—they are easy to oxidize, their cations are difficult to reduce. Copper metal is a relatively unreactive metal—it is harder to oxidize, but its cations are easy to reduce.

Teacher's Notes

Teacher Notes

5. *(Optional)* Consult a table of standard reduction potentials ($E°_{red}$): Determine the minimum voltage necessary for the electrolysis of aqueous potassium iodide.

Hint: $E°_{cell} = E°_{red}$ (cathode) − $E°_{red}$ (anode)

Oxidation (anode): $2I^-(aq) \rightarrow I_2(s) + 2e^-$ $E°_{red} = +0.54$ V

Reduction (cathode): $2H_2O(l) + 2e^- \rightarrow H_2(g) + 2OH^-(aq)$ $E°_{red} = -0.83$ V

$E°_{cell} = E°_{red}$ (cathode) − $E°_{red}$ (anode) = −0.83 V − 0.54 V = −1.37 V

The minimum cell voltage required for this nonspontaneous reaction is 1.37 V.

Note to teachers: *Remind students that standard reduction potentials are for reactions written as reductions. The anode half-reaction (oxidation) must be reversed when looking up the standard reduction potential. E° values are also based on 1 M solutions of all ions, which was not the case in this experiment.*

Electrolysis Reactions

Teacher's Notes

Supplementary Information

Teacher Notes

Pencil Electrolysis — A Combined Electrode–Battery Assembly

A very simple electrolysis device can be used to demonstrate a variety of electrochemical reactions.

Materials

Two sharpened #2 pencils
Battery, 9-V
Battery cap with alligator clip leads
Petri dish
Electrician's tape
Pocket knife

Construction

1. For each pencil, whittle away one side of the pencil shaft just below the eraser. This may be accomplished with a sharp pocket knife or a wood file. Enough of the shaft must be removed to expose the graphite as shown in the drawing. (Be careful not to break the graphite.)

2. Place the battery cap, with attached alligator clips, on the battery, then position the battery between the two pencils, just below the carved away sections, as shown at right. Wrap tape securely around the assembly. Now attach one clip to each pencil so that it is in good contact with the exposed graphite.

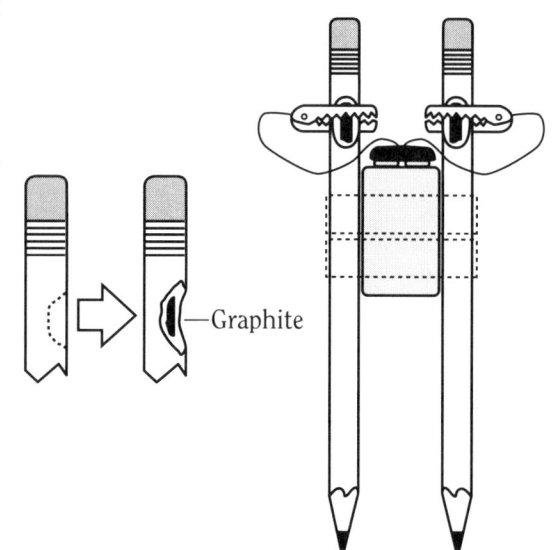

Presentation

Dissolve a few grams of sodium sulfate or other salt in 50 mL of water. Add an acid–base indicator such as bromthymol blue or universal indicator.

Put a Petri dish bottom on the overhead projector, and pour the salt solution into the dish.

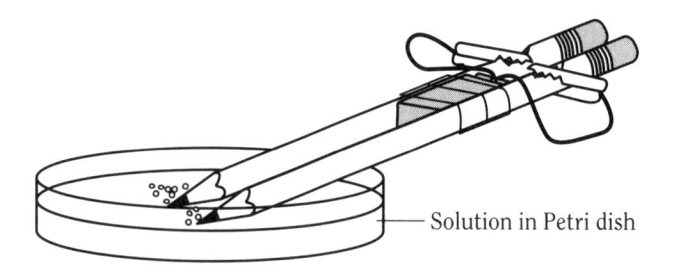

Holding the pencil-electrolysis device at an angle, place the two pencil tips in the solution. Observe. Bubbling should be apparent at both electrodes (pencil tips), about twice as vigorous at one tip as at the other. Color changes at each electrode indicate the formation of H^+ and OH^- ions. Ask students to write what reactions are occurring at each electrode.

Flinn ChemTopic™ Labs — Electrochemistry

Demonstrations

Microscale Electrolysis
The Rocket Reaction

Teacher Notes

Introduction

The decomposition of a compound into simpler substances by means of an electrical current is called electrolysis. When electricity is passed through a solution of water containing an electrolyte, hydrogen gas and oxygen gas are obtained. In this microscale demonstration, the gas mixture is collected in a pipet bulb by water displacement and then ignited with a spark. The resulting "bulb rocket" shoots across the room, proving that the gas mixture contains the 2:1 stoichiometric ratio of hydrogen and oxygen.

Concepts

- Electrolysis
- Oxidation–reduction
- Decomposition reaction
- Combustion reaction

Materials

Battery, 9-volt
Battery cap with alligator clip leads
Beakers, 600-mL, 2
Cardboard, 16 cm × 10 cm
Copper wire, insulated, 22 cm
Graphite pencil "lead," 0.9 mm
Paper towels
Pencil
Piezoelectric igniter

Pipets, Beral-type, extra large bulb, 7
Pipet, Beral-type, super jumbo
Sodium sulfate solution, Na_2SO_4, 1 M, 300 mL
Scissors
Test tube rack
Transparent tape
Thumbtack or pushpin
Water
Electrical tape

Safety Precautions

Exercise care when using the piezoelectric igniter—do not touch the bare copper wire. Wear chemical splash goggles, chemical-resistant gloves, and a chemical-resistant apron. Please review current Material Safety Data Sheets for additional safety, handling, and disposal information.

Preparation

Gas-Generator Bulb

1. Cut the stem off one super jumbo, Beral-type pipet, about one cm from the pipet bulb (Figure 1a). *Note:* One gas-generator bulb can be used to fill at least 10 "rockets."

2. Use a thumbtack or pushpin to make two small holes in the stem side of the pipet bulb. Place the holes as far apart as possible on opposite sides of the stem.

3. Break the pencil lead to obtain two, 2-cm pieces. Place one piece of pencil lead into each hole in the pipet bulb (Figure 1b, page 52). The pencil leads should fit snugly into the holes. Do not allow the electrodes to touch.

Use heavy-duty, 0.9 mm mechanical pencil lead for the electrodes. Thinner pencil leads (0.7 or 0.5 mm) are more fragile and may be more likely to break. Keep the holes in the gas generating pipet bulb as small as possible. The electrodes must fit snugly into the bulb so that the gases will not escape through the holes. The gas generator bulb may be reused many times (7–10) to collect fresh gas mixtures.

Demonstrations

Teacher Notes

Gas-Collecting Bulbs

4. Cut the stems off one or more (depending on how many rockets are desired) extra large, Beral-type pipets, about one cm from the pipet bulb (Figure 2, page 52).

Piezoelectric Igniter (Modified)

5. Cut two pieces of insulated copper wire, about 10 cm and 12 cm, respectively. Strip about 1.5 cm of insulation off the end of the shorter wire and about 1 cm off the end of the longer wire (Figure 3a, page 52).

6. Curl the 1.5 cm end of bare wire around a pencil tip to make a small, tight coil. Place the coil around the metal post in the center of the piezoelectric igniter (Figure 3b, page 52).

7. Line up the bare wire end of the second piece of insulated wire directly on top of the bare copper wire that runs down the side of the piezoelectric igniter (Figure 3c, page 52). Tape the stripped wire to the copper wire on the piezoelectric lighter in two places as shown.

8. Line up the the two pieces of insulated wire and tape them together just above the metal post (Figure 3d, page 52).

9. Cut the insulated ends of the two pieces of wire so that they are the same length. Tape the insulated ends together so that there is only a small "spark gap" between the wires (Figure 3e, page 52). This is the sparking portion of the modified igniter.

10. Wrap the body of the igniter with electrical tape for safety.

Rocket Launch Pad

11. Cut a 16 cm × 10 cm piece of thick cardboard and make a hole in the middle of the cardboard using a thumbtack. Widen the hole with the thumbtack until the hole is just large enough for the sparking wires of the modified piezoelectric igniter.

12. Tape one end of the cardboard to the table or benchtop and prop the cardboard up at an angle at the other end.

13. Place the wires of the piezoelectric igniter through the hole in the cardboard so that the sparking portion (step 9) will intrude past the stem portion of a gas-collecting (rocket) bulb placed over it (Figure 4, page 52).

Procedure

1. Completely fill the jumbo, gas-generator bulb with 1 M sodium sulfate solution. *Note:* A half-filled 600-mL beaker makes a convenient "filling station."

2. Place the gas-generator bulb in a test tube rack with some folded paper towels under the rack to catch the displaced water.

3. Fill an extra large, gas-collecting bulb with tap water.

4. Place the gas-collecting bulb over the stem of the gas generator.

5. Connect the 9-V battery to the pencil lead electrodes using a battery cap with alligator clip leads (Figure 5, page 52). Gas bubbles will be observed and the gas mixture will be collected by water displacement in the extra large bulb.

These directions were written for use of the piezoelectric igniter available from Flinn Scientific (Catalog No. AP6609).

Demonstrations

Teacher Notes

6. Collect the gas mixture until there is only a small amount of water remaining in the stem of the gas-collecting bulb. The water plug will prevent the gas from escaping prior to the "rocket launch." Disconnect the battery.

7. Carefully remove the gas-collecting bulb from the generator. Place the bulb over the wires from the piezoelectric igniter on the rocket launch pad. The "spark gap" should be above the water plug in the bulb (Figure 6, page 52).

8. Aim the rocket away from all spectators and then press the button on the piezoelectric igniter to ignite the gas mixture. The rocket will fly across the room with a loud bang!

9. Repeat as desired. The gas generator can be used 7–10 times. Use fresh rockets.

Disposal

Please consult your current *Flinn Scientific Catalog/Reference Manual* for general guidelines and specific procedures governing the disposal of laboratory waste. The electrolysis waste solution may be rinsed down the drain with excess water according to Flinn Suggested Disposal Method #26b.

Tips

- Compare the explosiveness of the gas mixture obtained by electrolysis versus an optimum 2:1 mixture of gases collected from separate hydrogen and oxygen gas generators. See the experiment "Micro Mole Rockets" in *Molar Relationships and Stoichiometry*, Volume 7 in the *Flinn ChemTopic™ Labs* series.

- To make an even bigger "splash" with this demonstration, use the following alternative procedure to ignite the gas mixture. Make a small hole near the bottom of a gas-collecting bulb with a thumbtack and widen the hole just enough to fit the wire ends of the modified piezoelectric igniter through. Position the igniter so that the "spark gap" is in the center of the bulb. Collect the gas mixture as described in steps 1–6 in the *Procedure* section. (Do not submerge more than the insulated portion of the wires when filling the gas-collecting bulb with water.) Holding the gas-filled bulb stem down, submerge the stem portion in a shallow pan of water. Carefully clamp the bulb in position and then ignite the gas mixture. With the "rocket" bulb firmly locked in place, the force of the resulting explosion produces a great big splash!

Discussion

Electrolysis of water is an oxidation–reduction reaction that occurs when an electrical current is passed through water containing an electrolyte (in this case, sodium sulfate). Reduction occurs at the cathode, where hydrogen gas is generated. Oxidation occurs at the anode, where oxygen gas is generated. Multiplying the reduction half-reaction by two (to balance electron transfer) and combining the two half-reactions gives the net equation for the decomposition of water. Notice that the equation for the electrolysis of water is the reverse of the combustion reaction of hydrogen. The latter reaction is the "rocket reaction" that is used to demonstrate the stoichiometric ratio of gases produced by electrolysis. The explosiveness of the rocket reaction confirms that the ratio of hydrogen gas to oxygen gas generated by electrolysis is indeed 2:1, the same ratio of hydrogen to oxygen in water.

Microscale Electrolysis

Demonstrations

Teacher Notes

Oxidation half-reaction (anode) $2H_2O \rightarrow O_2(g) + 4H^+ + 4e^-$

Reduction half-reaction (cathode) $2e^- + 2H_2O \rightarrow H_2(g) + 2OH^-$

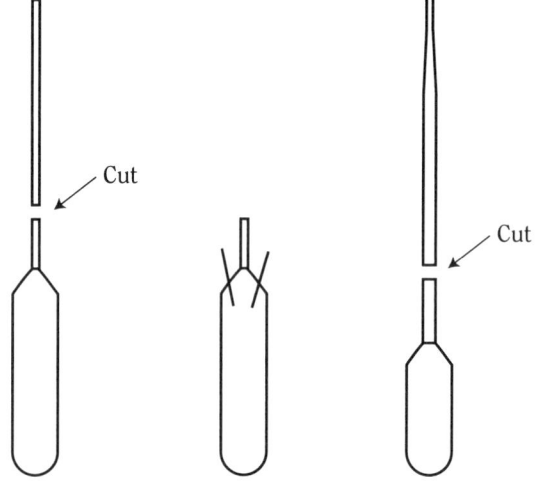

Decomposition of water $2H_2O \xrightarrow{\text{Electricity}} 2H_2 + O_2$

Combustion of hydrogen $2H_2 + O_2 \rightarrow 2H_2O + \text{Energy}$

Figure 1a.   **Figure 1b.**   **Figure 2.**

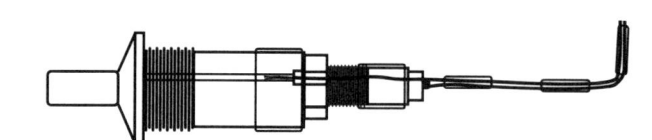

Figure 3e.
(Wrap with electrical tape when done.)

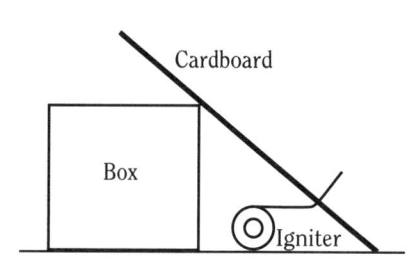

Figure 3a.

Figure 3b.

Figure 4.

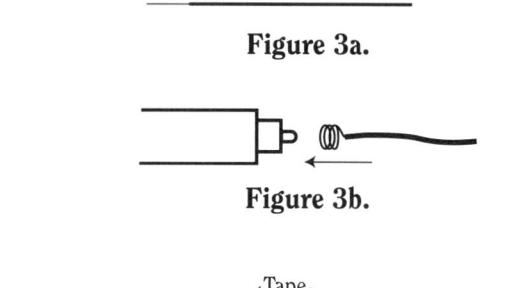

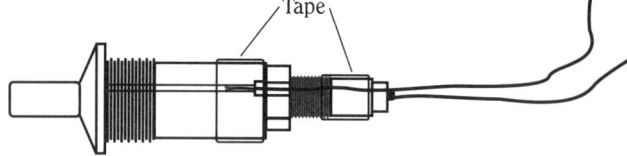

Figure 3c.

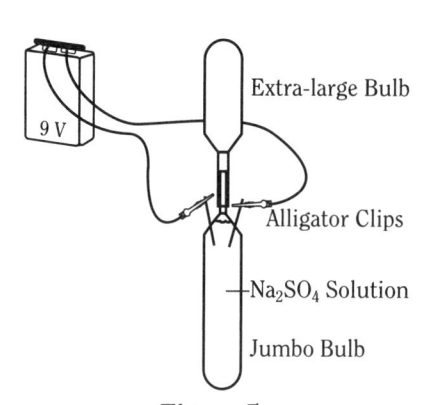

Figure 5.

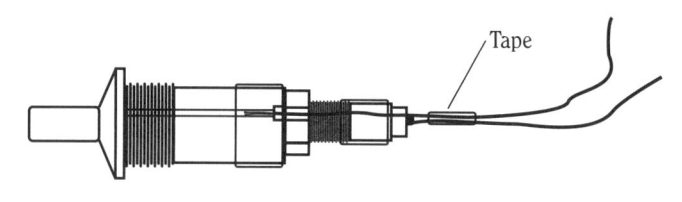

Figure 3d.

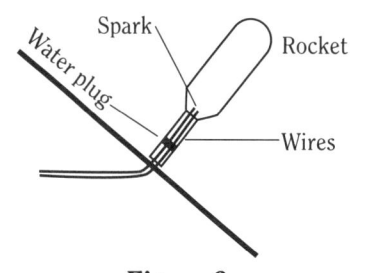

Figure 6.

Flinn ChemTopic™ Labs — Electrochemistry

Demonstrations

Teacher Notes

Hoffman Electrolysis
Color-Enhanced!

Introduction

The decomposition reaction of water using an electric current provides a great opportunity to illustrate fundamental principles of electrochemistry. What reaction occurs at the cathode versus the anode? What ions are produced at each electrode as the oxidation–reduction reaction proceeds? What is the relationship between the amount of electricity passed through the solution and the amount of gas obtained?

Concepts

- Electrolysis
- Anode vs. cathode
- Oxidation–reduction
- Current, coulombs, and electrons

Materials

Bromthymol blue indicator solution, 0.04%, 15 mL
Hydrochloric acid solution, HCl, 0.1 M, 10 mL
Sodium sulfate solution, Na_2SO_4, saturated, 200 mL
Sodium hydroxide solution, NaOH, 0.1 M, 10 mL
Sulfuric acid solution, H_2SO_4, 1 M, 200 mL (optional)

Ammeter (optional)	Barometer (optional)
Battery, 6-V, or power supply	Alligator clip leads
Beaker, 400- or 600-mL	Bunsen burner or matches
Graduated cylinder, 10-mL	Hoffman apparatus
Pipet, Beral-type	Power supply (optional)
Syringes, 10-mL, 2, with latex tubing	Test tubes, 15 × 125 mm, 2
Thermometer (optional)	Wood splints

Safety Precautions

Sulfuric acid solution is a corrosive liquid—avoid contact with eyes and skin. Do not operate a power supply with wet hands or in a wet area. Be sure the area is dry before turning on the power supply or closing the circuit. Wear chemical splash goggles, chemical-resistant gloves, and a chemical-resistant apron. Please review current Material Safety Data Sheets for additional safety, handling, and disposal information.

Preparation

1. Prepare the saturated sodium sulfate solution by dissolving 120 g of sodium sulfate decahydrate ($Na_2SO_4·10H_2O$) in 200 mL of deionized or distilled water.

2. Add 10 mL of bromthymol blue indicator solution to the saturated sodium sulfate solution. The resulting electrolysis solution should be a bright, "neutral" green color. If the solution is blue, add a small amount of 0.1 M hydrochloric acid to get the green color. If the solution is yellow, add a small amount of 0.1 M sodium hydroxide solution.

Hoffman Electrolysis

Demonstrations

3. Set up the Hoffman apparatus as shown in Figure 1. With the stopcocks open, pour the sodium sulfate–bromthymol blue electrolysis solution into the central bulb reservoir to fill the apparatus. Close the stopcocks when both "arms" are completely filled with solution.

4. *(Optional)* Demonstrate the acid–base color changes for bromthymol blue indicator. Add 5 drops of bromthymol indicator solution to 5 mL of 0.1 M hydrochloric acid solution in a test tube. *(The solution will turn yellow.)* Add a small amount of bromthymol indicator to 5 mL of 0.1 M sodium hydroxide solution in a second test tube. *(The solution will turn blue.)*

Figure 1. Hoffman Apparatus.

Procedure

Part A. Electrolysis in Living Color

1. Connect the platinum electrodes in the Hoffman apparatus to a 6-V battery or power supply. Attach the left-hand electrode to the positive terminal and the right-hand electrode to the negative terminal. *(In an electrolytic cell, the positive electrode is the anode and the negative electrode is the cathode.)*

2. Observe the immediate production of gas bubbles and the color changes in each arm of the Hoffman apparatus.

3. Compare the amount of gas generated at each electrode and identify each gas. *[The amount of oxygen is one-half the amount of hydrogen. Oxygen is obtained at the positive electrode (anode), hydrogen at the negative electrode (cathode).]*

4. Compare the indicator color changes at the anode versus the cathode. Identify the ions produced at each electrode. *(The color of the solution changes to yellow at the anode—the production of oxygen gas is accompanied by the formation of H^+ ions. The color of the solution changes to blue at the cathode—the production of hydrogen gas is accompanied by the formation of OH^- ions.)*

5. Disconnect the battery or turn off the power supply to stop the electrolysis. Using a syringe attached to the stopcock via a short length of latex tubing, withdraw a sample of each gas and test its properties. *("Squirting" oxygen gas onto a glowing wood splint reignites the flame. Hydrogen gas extinguishes the flame on a burning wood splint and produces a loud "pop.")*

6. Pour the electrolysis solution into a beaker and observe the final indicator color of the mixed solution. What happened to the H^+ and OH^- ions produced during electrolysis? *(The mixed electrolysis solution is green. The total number of H^+ ions produced during electrolysis is equal to the total number of OH^- ions and the ions exactly neutralize each other, producing water molecules.)*

Teacher Notes

Use the worksheet on page 56 to engage the students in this demonstration. The worksheet questions correspond to the rhetorical questions incorporated into the Procedure.

Demonstrations

Teacher Notes

Part B. Quantitative Electrolysis

7. *(Optional)* Fill the Hoffman apparatus with 1 M sulfuric acid solution and connect an ammeter in series with the power supply. Adjust the voltage to obtain a stable current (about 0.3 A) and record the initial time and the current.

8. Run the electrolysis until about 15 mL of hydrogen gas has been collected. Disconnect the power supply and record the time. Measure and record the exact amount of hydrogen gas produced. Measure and record the barometric pressure and the temperature of the electrolysis solution. *(Sample results are summarized below.)*

Barometric pressure	744 mm Hg
Electrolysis time	360 sec
Average current	0.30 A
Volume of hydrogen	14.55 mL
Temperature of solution	30.7 °C
Vapor pressure of water*	33 mm Hg
Moles of hydrogen produced	5.48×10^{-4} moles
Moles of electrons "consumed"	1.10×10^{-3} moles
Faraday constant (calculated)	98,200 C/mole
Percent error in Faraday constant	2%

*Literature value.

Disposal

Please consult your current *Flinn Scientific Catalog/Reference Manual* for general guidelines and specific procedures governing the disposal of laboratory waste. The sodium sulfate waste solution may be disposed of with excess water according to Flinn Suggested Disposal Method #26b. Sulfuric acid may be neutralized with base according to Flinn Suggested Disposal Method #24b.

Tips

- Bromthymol blue is yellow in acidic solution (excess H^+ ions) and blue in basic solutions (excess OH^- ions). It is green in neutral solutions (pH = 6.0–7.6).

- See the experiment "Introduction to Electrochemistry" and the demonstration "Microscale Electrolysis" for additional background information. Use the worksheet on the following page to engage students in the discussion. The answers to the worksheet questions have been incorporated into the obervations and results in the *Procedure* section.

- See "Quantitative Electrochemistry" for a discussion of the quantitative principles of electrochemistry and sample calculations involving current, coulombs, and electrons.

Calculations for Part B:

Pressure of hydrogen
744–33 = 711 mm Hg
(0.936 atm)

Moles of hydrogen
$$\frac{(0.936 \text{ atm})(0.0146 \text{ L})}{\left(\frac{.0821 \text{ L·atm}}{\text{mole K}}\right)(304 \text{ K})}$$
5.48×10^{-4} moles

Mole ratio
$$\frac{2 \text{ moles electrons}}{1 \text{ mole } H_2}$$

Coulombs
$$\frac{0.30 \text{ A} \times 1 \text{C/sec} \times 360 \text{ sec}}{1 \text{ A}}$$
108 C

Faraday constant
$$\frac{108 \text{ C}}{1.10 \times 10^{-3} \text{ moles}}$$
98,200 C/moles

Hoffman Electrolysis

Demonstrations

Name: _____

Class/Lab Period: _____

Hoffman Electrolysis Worksheet

1. What is the initial indicator color of the sodium sulfate electrolysis solution? What is the approximate pH of this solution?

2. Describe the observations that indicate a chemical reaction takes place during electrolysis.

3. Write the balanced chemical equation for the decomposition reaction of water to its elements.

4. Compare the amount of gas generated at each electrode. Why are the gases produced in different amounts? Identify the gas produced at the positive and the negative electrode, respectively.

5. Compare the color changes observed at the positive and negative electrodes. What ions are produced at each electrode? Explain.

6. Balance the following oxidation and reduction half-reactions for the decomposition of water. Hint: Hydrogen ions (H⁺) and hydroxide ions (OH⁻) are required to balance mass and charge.

   ☐ H_2O → O_2 + ☐ H^+ + ☐ e^-

   ☐ H_2O + ☐ e^- → $2H_2$ + ☐ OH^-

7. What is the final indicator color and pH of the mixed sodium sulfate and bromthymol blue solution after electrolysis?

8. What happened to the H⁺ and OH⁻ ions produced during electrolysis? Explain how the oxidation and reduction half-reactions may be combined to give the balanced chemical equation for the decomposition of water.

Teacher Notes

See the experiment "Introduction to Electrochemistry" in this book for answers to these worksheet questions.

Demonstrations

Teacher Notes

The Tin Man
Tin(IV), Tin(II), and Tin(0)

Introduction

What happens in an electrolytic cell if one of the atoms or ions in a compound may be both oxidized and reduced? The electrolysis of tin(II) chloride provides a stunning example—the resulting tin crystals are beautiful to behold.

Concepts

- Electrolysis
- Oxidation–reduction
- Anode vs. cathode

Materials

Copper wire, 1–2 cm (optional)

Tin(II) chloride solution, $SnCl_2$,* 1 M in HCl, 25 mL

Overhead projector

Battery, 9-V

Battery cap with alligator clip leads

Paper clips, small, 2

Petri dish

*Also called stannous chloride.

Safety Precautions

The acidic tin(II) chloride solution is corrosive to body tissue and moderately toxic by ingestion. Avoid contact of all chemicals with eyes and skin. Wear chemical splash goggles, chemical-resistant gloves, and chemical-resistant apron. Please review current Material Safety Data Sheets for additional safety, handling, and disposal information.

Preparation

Prepare 1 M tin(II) chloride solution by dissolving 22.5 g of tin(II) chloride dihydrate ($SnCl_2 \cdot 2H_2O$) in 100 mL of 1 M hydrochloric acid solution.

Procedure

1. Place a Petri dish on an overhead projector and turn on the projector.

2. Add about 25 mL of tin(II) chloride solution to the Petri dish until the dish is one-third to one-half full.

3. Connect one paper clip to each alligator clip lead on the battery cap and clip the paper clips to opposite sides of the Petri dish.

4. Attach the battery cap with alligator clip leads to the 9-V battery and observe the changes at the positive electrode (the anode) and the negative electrode (the cathode). *[A milky-white precipitate of tin(IV) chloride appears at the anode and beautiful metallic tin(0) crystals form at the cathode.]*

See the experiment "Electrolysis Reactions" for additional examples of oxidation–reduction reactions in the electrolysis of aqueous salt solutions.

The Tin Man

Demonstrations

5. Allow the reaction to continue for 30–60 seconds and observe the pattern of crystal growth. *[The tin crystals grow out from the cathode in a feather-like (fractal) pattern, producing a beautiful "crystal tree." See Figure 1.]*

6. Gently remove the alligator clip leads from the paper clips and switch the polarity of the electrodes. Attach the alligator clip from the negative battery post to the previous anode (left-hand side above) and the alligator clip from the positive battery post to the previous cathode (the right-hand side above.)

7. Observe the changes at the new anode and the new cathode. *(The reactions are reversed—new tin crystals grow from the previous anode and the tin crystals at the previous cathode dissolve into solution.)*

8. Disconnect the battery to stop the reaction.

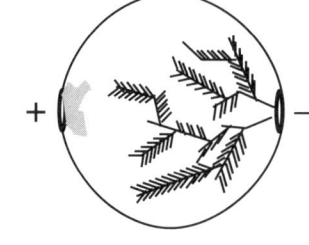

Figure 1.

Teacher Notes

Disposal

Please consult your current *Flinn Scientific Catalog/Reference Manual* for general guidelines and specific procedures governing the disposal of laboratory waste. Decant the liquid from the Petri dish. The solution may be neutralized with base according to Flinn Suggested Disposal Method #24b. The solids may be disposed of in the trash according to Flinn Suggested Disposal Method #26a.

Tips

- A black coating forms on the paper clip electrodes when the paper clips are placed in the tin(II) chloride solution. This is due to a spontaneous single replacement reaction between the metal in the paper clips and tin(II) ions. The black coating will not interfere with the electrolysis reaction.

- Place a small piece of copper wire in the center of the Petri dish. Position the copper wire so that its ends are pointing toward opposite electrodes. During electrolysis, a Sn(IV) precipitate forms at the anode and Sn(0) crystals grow at the cathode. The copper wire acts as a second set of electrodes—the pattern is repeated.

Discussion

Electrolysis is defined as the decomposition of a compound using an electric current. An electrolytic cell consists of an anode, a cathode, a conducting solution, and a power source. Oxidation occurs at the anode, reduction at the cathode. In this demonstration, an electric current causes an oxidation–reduction reaction to occur. Tin(II) ions are oxidized to tin(IV) ions at the anode and are reduced to metallic tin at the cathode. Insoluble tin(IV) chloride is observed as a milky white precipitate at the anode. The overall reaction is a disproportionation reaction.

Oxidation half-reaction (anode) $\quad Sn^{2+}(aq) \rightarrow Sn^{4+}(aq) + 2e^-$

$\qquad\qquad\qquad\qquad\qquad\qquad\quad Sn^{4+}(aq) + 4Cl^-(aq) \rightarrow SnCl_4(s)$

Reduction half-reaction (cathode) $\quad Sn^{2+}(aq) + 2e^- \rightarrow Sn(s)$

Overall reaction (disproportionation) $\quad 2Sn^{2+}(aq) + 4Cl^-(aq) \rightarrow SnCl_4(s) + Sn(s)$

Flinn ChemTopic™ Labs — Electrochemistry

Demonstrations

Teacher Notes

Orange Juice Clock
Electricity from a Chemical Reaction

Introduction

Build an electrochemical cell using orange juice and copper and magnesium metal electrodes. The resulting spontaneous oxidation–reduction reaction will generate enough electricity to power a battery-operated clock. Why does the clock work? How long will the clock run? What happens if the metals are reversed?

Concepts

- Electrochemistry
- Cell potential
- Metal activity
- Anode vs. cathode

Materials

Battery-operated clock
Beaker, 600-mL
Connector cords with alligator clips, 2
Copper foil or sheet, 5 mm × 30 cm
Magnesium ribbon, 30 cm
Orange juice or soda, 400 mL
Pencil
Support (ring) stand and ring
Utility clamp

Safety Precautions

Magnesium is a flammable solid. Avoid contact with flames and heat. Any food-grade items that have been brought into the lab are considered laboratory chemicals and are for lab use only. Do not taste or ingest any materials in the laboratory and do not remove any remaining food items after they have been used in the lab. Wear chemical splash goggles, chemical-resistant gloves, and chemical-resistant apron. Please review current Material Safety Data Sheets for additional safety, handling, and disposal information.

Procedure

1. Coil one end of the magnesium ribbon around a pencil to form a loose coil about 7 mm in diameter and 5 cm long.

2. Attach the coiled magnesium ribbon to the *negative* terminal of a battery-operated wall clock using a connector cord with alligator clips on both ends.

3. In a similar manner, coil the copper strip and attach it to the *positive* terminal of the clock.

4. Place about 400 mL of orange juice or soda in a 600-mL beaker on the base of a support stand.

5. Attach the clock to the support stand using a ring or a utility clamp and position the clock so that the metal electrodes will fit easily into the orange juice when it is time to start the demonstration (and the clock).

6. Set the time on the clock to correspond with the time the students will arrive.

See the "Electrochemical Clock Kit" available from Flinn Scientific (Catalog No. AP8718) for a wonderful inquiry-based activity using the orange juice clock.

Demonstrations

7. To start the demonstration, immerse the metal electrodes into the orange juice and ask students to record their observations. *(The clock will start running and the orange juice will begin to bubble, froth, and foam.)*

8. At this point, the students will naturally start to ask questions. Use the students' questions to start an open-ended discussion of the reactions that are occurring, how the electricity is generated, etc.

Disposal

Please consult your current *Flinn Scientific Catalog/Reference Manual* for general guidelines and specific procedures governing the disposal of laboratory waste. The magnesium ribbon should be left in the orange juice (or a similar acidic solution, such as 1 M hydrochloric acid) until it has completely dissolved. The orange juice waste solution may then be rinsed down the drain with excess water according to Flinn Suggested Disposal Method #26b.

Tips

- For best results, polish the magnesium and copper metal with sandpaper or steel wool before use. Rinse well with distilled water and pat dry. Alternatively, the metals may be cleaned by dipping them (briefly, in the case of magnesium) into 1 M acetic acid.

- See "Observations on Lemon Cells," by Jerry Goodisman, in the *Journal of Chemical Education*, Volume 78, No. 4, pp 516–518 (April 2001), for a thorough explanation of the net redox reaction and the cell potential for the orange juice electrochemical cell.

Discussion

The magnesium electrode is the *anode* (negative electrode) in the orange juice electrochemical cell—magnesium metal is oxidized to Mg^{2+} ions. The copper electrode is the *cathode* (positive electrode). The copper electrode is an inert electrode—hydrogen (H^+) ions from the citric acid solution are reduced to hydrogen gas (H_2) on the surface of the electrode.

Oxidation half-reaction (anode)	$Mg(s) \rightarrow Mg^{2+}(aq) + 2e^-$
Reduction half-reaction (cathode)	$2H^+(aq) + 2e^- \rightarrow H_2(g)$
Net reaction	$Mg(s) + 2H^+(aq) \rightarrow Mg^{2+}(aq) + H_2(g)$

The magnesium ribbon dissolves over the course of the electrochemical reaction. The reaction is spontaneous and the cell is irreversible. Copper has a higher electrical potential than magnesium. When the metal conductors are connected by means of an external wire circuit (the clock), there is a positive current flow from the copper to the magnesium. *Electrons flow in the opposite direction (from the magnesium to the copper)*. Within the electrochemical cell, ions carry the current through the solution. Anions move toward the anode, cations toward the cathode, to prevent charge buildup at the electrodes. The clock will not run if the polarity of the magnesium and copper electrodes is switched. (The clock may run backwards, however, if there are no diodes in the circuit.)

Teacher Notes

The orange juice clock is powered by a spontaneous oxidation–reduction reaction. When the electrodes are incorporated into a closed electrical circuit, electrons flow through the external wire. An electrochemical cell that generates electricity is called a voltaic cell. In a voltaic cell, the anode is the negative electrode and the cathode is the positive electrode.

Teacher Notes

Basic Electrophoresis
Migration of Ions in an Electric Field

Introduction

One of the most poorly understood concepts in electrochemistry has to do with the flow of electricity in an electrochemical cell. Many students assume that the flow of electricity through the solution(s) is due to the flow of electrons. In all types of electrochemical cells, electrons carry the current through an external wire or conductor, but ions carry the current through the solution. Anions move toward the anode, cations toward the cathode, to prevent charge buildup at the electrodes. A central feature of electrochemical cells, migration of ions in an electric field is also the basic principle in electrophoresis.

Concepts

- Electrophoresis
- Migration of ions
- Positive and negative electrodes

Materials

Agarose, electrophoresis-grade, 0.8 g

Buffer, pH 10, 500 mL

Methylene blue, 0.1% in pH 10 buffer, 50 mL

Phenol red, 0.1% in pH 10 buffer, 50 mL

Sucrose, 50 g

Electrophoresis apparatus, mini-gel (includes casting tray, rubber gaskets, and combs)

Micropipet, 10-μL, or needle-tip disposable pipet

Power source for electrophoresis apparatus (includes grounded wire cords)

Thermometer

Safety Precautions

Do not operate the power source with wet hands or in a wet area. Make sure the power supply is off before connecting the leads to the electrophoresis apparatus. Turn off the power supply before disconnecting the leads and removing the cover at the end of the demonstration. Wear chemical splash goggles, chemical-resistant gloves, and chemical-resistant apron. Please review current Material Safety Data Sheets for additional safety, handling, and disposal information.

Procedure

1. Place the rubber gaskets on the ends of the mini-gel casting tray and insert a 6-well comb in the *middle* of the tray.

2. Fill the electrophoresis chamber with about 100 mL of pH 10 buffer solution.

3. Prepare a 0.8% agarose gel by dissolving 0.4 g of agarose in 50 mL of pH 10 buffer solution at about 80 °C. Cool the agarose gel solution to about 50 °C, and then pour the solution into the gel casting tray.

Demonstrations

Teacher Notes

4. After the gel has solidified (about 20 minutes), remove the gaskets from the end of the tray and submerge the gel (on the tray) in the buffer in the electrophoresis chamber.

5. Prepare 0.1% solutions of methylene blue and phenol red: Dissolve 0.05 g of the dye in 50 mL of pH 10 buffer and add 25 g of sucrose to each dye solution.

6. Using a separate pipet tip for each dye solution, inject 10 μL of the methylene blue gel-loading solution into well #2, 10 μL of a 1:1 mixture of methylene blue and phenol red into well #4, and 10 μL of phenol red gel-loading solution into well #6.

7. Place the safety cover on the electrophoresis chamber. Make sure the power is off before connecting the power leads to the power source. Select the desired voltage (70 V) and turn on the power source. Run the electrophoresis 15–20 minutes or until the dyes have migrated to about 1 cm from the end of the gel in either direction.

8. Observe the direction of ion migration for each dye.

Disposal

Please consult your current *Flinn Scientific Catalog/Reference Manual* for general guidelines and specific procedures governing the disposal of laboratory waste. The gel may be disposed of in the trash according to Flinn Suggested Disposal Method #26a. The electrophoresis buffer solution may be reused several times before being disposed of down the drain with excess water according to Flinn Suggested Disposal Method #26b.

Tips

- The pH 10 buffer solution may be conveniently prepared using a Chemvelopes® buffer envelope (Catalog No. B0120) or buffer capsules (Catalog No. B0109).

- Sucrose is added to the dye solutions (step 5) to increase the density and prepare a "gel-loading" solution. The dense gel-loading solution will remain in the well.

- Typical protein or DNA electrophoresis gels can be stored in plastic bags with a small amount of buffer solution. The gel in this demonstration, however, cannot be stored, because the low-molecular weight dyes slowly diffuse through the gel.

- Many organic dyes and acid–base indicators may be used. Try bromphenol blue, crystal violet, malachite green, etc.

Discussion

Methylene blue is an ionic compound containing a colored, positively charged organic dye molecule and a chloride counterion. Phenol red is a colored organic dye that acts as an acid–base indicator. At pH 10, phenol red carries a double negative charge. Methylene blue (turquoise band) migrates toward the negative electrode during electrophoresis, while phenol red (bright pink band) migrates toward the positive electrode. The mixture of dyes (well #4), which was purple to start, separates into two bands, turquoise and bright pink, which migrate to opposite electrodes during electrophoresis.

Demonstrations

Teacher Notes

Lemon Battery Contest
Metal Activity and Cell Potentials

Introduction
Design a battery using metal strips and a lemon or an orange. Good luck!

Concepts
- Electrochemistry
- Metal activity
- Cell potential
- Anode vs. cathode

Materials
Connector cords with alligator clips, 4
Digital voltmeter (multimeter) or voltage sensor
Lemon, grapefruit or orange
Metal strips (aluminum, copper, iron, magnesium, tin, and zinc), 1 cm × 4 cm, 2 each
Sandpaper or steel wool
Computer interface system, such as LabPro or CBL (optional)
Computer or calculator for data collection (optional)

Safety Precautions
Magnesium is a flammable solid. Avoid contact with flames and heat. Any food-grade items that have been brought into the lab are considered laboratory chemicals and are for lab use only. Do not taste or ingest any materials in the laboratory and do not remove any remaining food items after they have been used in the lab. Wear chemical splash goggles, chemical-resistant gloves, and chemical-resistant apron. Please review current Material Safety Data Sheets for additional safety, handling, and disposal information.

Procedure

1. Cut strips of metal sheet (see the *Materials* list), about 1 cm wide by 4 cm long. Place the metal strips on a labeled sheet of paper to keep track of their identity.

2. "Prime" the lemon or orange by rolling it briskly on the table or benchtop to soften the skin.

3. Insert two different metal strips about 1–2 cm apart into the lemon or orange. The metals should penetrate the fruit to a depth of at least 2–3 cm.

4. Attach a voltage lead (or connector cord with alligator clips) from the multimeter or voltage sensor to each metal electrode.

5. Measure the voltage. *If a positive voltage reading is obtained,* record the voltage and note which metal is attached to the positive lead and which is attached to the negative lead. *Note:* If a negative voltage reading is obtained, reverse the polarity of the metal electrodes to obtain a positive reading.

6. Observe any signs of a chemical reaction in or around each metal electrode.

The "Metal Electrode Set" available from Flinn Scientific (Catalog No. AP4602) contains six metal electrodes (aluminum, brass, copper, iron, lead, and zinc) that are perfect for this activity.

Demonstrations

7. Repeat steps 3–6 to measure the voltage for different combinations of metals. Remember to record the identity of the positive and the negative electrodes when a positive voltage is obtained. What combination of metals gives the highest voltage?

8. *(Optional)* Can pairs of metals be connected in series to increase the voltage of a lemon battery?

9. *(Optional)* Does the voltage depend on the separation between the metals?

10. *(Optional)* Is the battery voltage stable over time? How long will the battery last?

Teacher Notes

Disposal

Please consult your current *Flinn Scientific Catalog/Reference Manual* for general guidelines and specific procedures governing the disposal of laboratory waste. Inert metals such as copper or tin may be cleaned and saved for future use. Metals that show signs of chemical reaction may be disposed of in the solid trash according to Flinn Suggested Disposal Method #26a.

Tips

- For best results, polish the metal strips with sandpaper or steel wool before use. Rinse well with distilled water and pat dry.

- See "Observations on Lemon Cells," by Jerry Goodisman, in the *Journal of Chemical Education,* Volume 78, No. 4, pp 516–518 (April 2001), for an explanation of the results obtained with typical electrochemical cells in lemons and orange juice.

Sample Results

Positive Electrode (Cathode)	Negative Electrode (Anode)	Voltage
Cu	Mg	1.9 V
Cu	Zn	1.0 V
Cu	Al	0.8 V
Cu	Fe	0.5 V
Cu	Sn	0.5 V
Sn	Mg	1.3 V
Sn	Zn	0.5 V
Sn	Al	0.1 V
Sn	Fe	0.0 V
Fe	Mg	1.3 V
Fe	Zn	0.5 V
Fe	Al	0.2 V
Al	Mg	1.2 V
Al	Zn	0.3 V
Zn	Mg	1.0 V

Flinn ChemTopic™ Labs — Electrochemistry

Demonstrations

Teacher Notes

Discussion

When two metals having different electrical potentials are placed in a solution of an electrolyte (such as a lemon or an orange, with its high concentration of citric acid), a crude electrochemical cell is set up. The cell is a *voltaic cell* resulting from a spontaneous oxidation–reduction reaction. The open-circuit potential difference between the reactions at each electrode is the cell potential, which can be measured with a digital voltmeter or a voltage sensor. The more active metal in a pair is *usually* the negative electrode (the anode). Magnesium, for example, which is the most active metal that was tested, always appears as the anode. (Aluminum is a significant exception to this general trend.) Significant bubbling is observed when magnesium is used as the electrode.

In general, the lemon-cell potential is greater when the two metals are further apart in the electrochemical series. The cell potential does *not* correlate well, however, with the difference in standard reduction potentials between the two metals. The voltage of a lemon cell battery decreases over time, suggesting that the overall reaction in the cell is irreversible.

The highest voltage (+1.9 V) in this demonstration was obtained with the copper–magnesium lemon cell with copper as the positive electrode (the cathode) and magnesium as the negative electrode (the anode). The most likely reactions in this cell are oxidation of magnesium to Mg^{2+} ions at the anode and reduction of H^+ ions to hydrogen gas at the cathode. (The citric acid content in a lemon or an orange is a rich source of H^+ ions.) The copper electrode is an inert electrode in this cell.

Oxidation half-reaction (anode) $\quad\quad Mg(s) \rightarrow Mg^{2+}(aq) + 2e^-$

Reduction half-reaction (cathode) $\quad\quad 2H^+(aq) + 2e^- \rightarrow H_2(g)$

Net reaction $\quad\quad Mg(s) + 2H^+(aq) \rightarrow Mg^{2+}(aq) + H_2(g)$

Lemon Battery Contest

Safety and Disposal

Safety and Disposal Guidelines

Safety Guidelines

Teachers owe their students a duty of care to protect them from harm and to take reasonable precautions to prevent accidents from occurring. A teacher's duty of care includes the following:

- Supervising students in the classroom.
- Providing adequate instructions for students to perform the tasks required of them.
- Warning students of the possible dangers involved in performing the activity.
- Providing safe facilities and equipment for the performance of the activity.
- Maintaining laboratory equipment in proper working order.

Safety Contract

The first step in creating a safe laboratory environment is to develop a safety contract that describes the rules of the laboratory for your students. Before a student ever sets foot in a laboratory, the safety contract should be reviewed and then signed by the student and a parent or guardian. Please contact Flinn Scientific at 800-452-1261 or visit the Flinn Website at www.flinnsci.com to request a free copy of the Flinn Scientific Safety Contract.

To fulfill your duty of care, observe the following guidelines:

1. **Be prepared.** Practice all experiments and demonstrations beforehand. Never perform a lab activity if you have not tested it, if you do not understand it, or if you do not have the resources to perform it safely.

2. **Set a good example.** The teacher is the most visible and important role model. Wear your safety goggles whenever you are working in the lab, even (or especially) when class is not in session. Students learn from your good example—whether you are preparing reagents, testing a procedure, or performing a demonstration.

3. **Maintain a safe lab environment.** Provide high-quality goggles that offer adequate protection and are comfortable to wear. Make sure there is proper safety equipment in the laboratory and that it is maintained in good working order. Inspect all safety equipment on a regular basis to ensure its readiness.

4. **Start with safety.** Incorporate safety into each laboratory exercise. Begin each lab period with a discussion of the properties of the chemicals or procedures used in the experiment and any special precautions—including goggle use—that must be observed. Pre-lab assignments are an ideal mechanism to ensure that students are prepared for lab and understand the safety precautions. Record all safety instruction in your lesson plan.

5. **Proper instruction.** Demonstrate new or unusual laboratory procedures before every activity. Instruct students on the safe way to handle chemicals, glassware, and equipment.

Safety and Disposal

6. **Supervision.** Never leave students unattended—always provide adequate supervision. Work with school administrators to make sure that class size does not exceed the capacity of the room or your ability to maintain a safe lab environment. Be prepared and alert to what students are doing so that you can prevent accidents before they happen.

7. **Understand your resources.** Know yourself, your students, and your resources. Use discretion in choosing experiments and demonstrations that match your background and fit within the knowledge and skill level of your students and the resources of your classroom. You are the best judge of what will work or not. Do not perform any activities that you feel are unsafe, that you are uncomfortable performing, or that you do not have the proper equipment for.

Safety Precautions

Specific safety precautions have been written for every experiment and demonstration in this book. The safety information describes the hazardous nature of each chemical and the specific precautions that must be followed to avoid exposure or accidents. The safety section also alerts you to potential dangers in the procedure or techniques. Regardless of what lab program you use, it is important to maintain a library of current Material Safety Data Sheets for all chemicals in your inventory. Please consult current MSDS for additional safety, handling, and disposal information.

Disposal Procedures

The disposal procedures included in this book are based on the Suggested Laboratory Chemical Disposal Procedures found in the *Flinn Scientific Catalog/Reference Manual*. The disposal procedures are only suggestions—do not use these procedures without first consulting with your local government regulatory officials.

Many of the experiments and demonstrations produce small volumes of aqueous solutions that can be flushed down the drain with excess water. Do not use this procedure if your drains empty into groundwater through a septic system or into a storm sewer. Local regulations may be more strict on drain disposal than the practices suggested in this book and in the *Flinn Scientific Catalog/Reference Manual*. You must determine what types of disposal procedures are permitted in your area—contact your local authorities.

Any suggested disposal method that includes "discard in the trash" requires your active attention and involvement. Make sure that the material is no longer reactive, is placed in a suitable container (plastic bag or bottle), and is in accordance with local landfill regulations. Please do not inadvertently perform any extra "demonstrations" due to unpredictable chemical reactions occurring in your trash can. Think before you throw!

Finally, please read all the narratives before you attempt any Suggested Laboratory Chemical Disposal Procedure found in your current *Flinn Scientific Catalog/Reference Manual*.

Flinn Scientific is your most trusted and reliable source of reference, safety, and disposal information for all chemicals used in the high school science lab. To request a complimentary copy of the most recent *Flinn Scientific Catalog/Reference Manual,* call us at 800-452-1261 or visit our Web site at www.flinnsci.com.

National Science Education Standards

Experiments and Demonstrations

Content Standards	Introduction to Electrochemistry	Measuring Cell Potentials	Quantitative Electrochemistry	Electrolysis Reactions	Microscale Electrolysis	Hoffman Electrolysis	The Tin Man	Orange Juice Clock	Basic Electrophoresis	Lemon Battery Contest
Unifying Concepts and Processes										
Systems, order, and organization										
Evidence, models, and explanation	✓	✓		✓	✓	✓		✓	✓	✓
Constancy, change, and measurement		✓	✓			✓				
Evolution and equilibrium										
Form and function										
Science as Inquiry										
Identify questions and concepts that guide scientific investigation	✓	✓	✓	✓	✓	✓	✓	✓	✓	✓
Design and conduct scientific investigations	✓	✓	✓	✓	✓	✓		✓		✓
Use technology and mathematics to improve scientific investigations		✓	✓			✓				✓
Formulate and revise scientific explanations and models using logic and evidence	✓	✓		✓				✓		✓
Recognize and analyze alternative explanations and models										
Communicate and defend a scientific argument										
Understand scientific inquiry	✓	✓	✓	✓						✓
Physical Science										
Structure of atoms										
Structure and properties of matter		✓		✓			✓	✓	✓	✓
Chemical reactions	✓	✓	✓	✓	✓	✓	✓	✓		✓
Motions and forces	✓								✓	
Conservation of energy and the increase in disorder										
Interactions of energy and matter	✓	✓	✓	✓	✓	✓	✓	✓	✓	✓

National Science Education Standards

Experiments and Demonstrations

Content Standards (continued)	Introduction to Electrochemistry	Measuring Cell Potentials	Quantitative Electrochemistry	Electrolysis Reactions	Microscale Electrolysis	Hoffman Electrolysis	The Tin Man	Orange Juice Clock	Basic Electrophoresis	Lemon Battery Contest
Science and Technology										
Identify a problem or design an opportunity										✓
Propose designs and choose between alternative solutions										✓
Implement a proposed solution										
Evaluate the solution and its consequences										
Communicate the problem, process, and solution										
Understand science and technology		✓	✓					✓	✓	✓
Science in Personal and Social Perspectives										
Personal and community health										
Population growth										
Natural resources										
Environmental quality										
Natural and human-induced hazards										
Science and technology in local, national, and global challenges										
History and Nature of Science										
Science as a human endeavor		✓								✓
Nature of scientific knowledge										
Historical perspectives	✓		✓			✓	✓		✓	

Master Materials Guide

(for a class of 30 students working in pairs)

Experiments and Demonstrations

Chemicals	Flinn Scientific Catalog No.	Introduction to Electrochemistry	Measuring Cell Potentials	Quantitative Electrochemistry	Electrolysis Reactions	Microscale Electrolysis	Hoffman Electrolysis	The Tin Man	Orange Juice Clock	Basic Electrophoresis	Lemon Battery Contest
Agarose, electrophoresis grade	A0132									1 g	
Bromthymol blue solution, 0.04%	B0173	50 mL				15 mL					
Buffer envelopes, pH 10	B0120									1	
Copper strips	C0182		15 cm	30							
Copper wire, 18-gauge	C0148								2 cm		
Copper(II) bromide	C0210				9 g						
Copper(II) sulfate solution, 1 M	C0246		15 mL	1.5 L							
Hydrochloric acid solution, 1 M	H0013						100 mL				
Hydrochloric acid solution, 0.1 M	H0014					10 mL					
Iron strips	I0058	15 cm									
Iron(II) sulfate heptahydrate	F0016	14 g									
Isopropyl alcohol	I0020		600 mL								
Magnesium ribbon	M0139	15 cm									30 cm
Magnesium sulfate heptahydrate	M0016	13 g									
Methylene blue	M0072										1 g
Phenolphthalein solution, 0.5%	P0115				15 mL						
Phenol red	P0097									1 g	
Potassium iodide solution, 0.5 M	P0171				150 mL						
Sandpaper	S0165	1									1
Silver foil	S0270	15 cm									
Silver nitrate solution, 1 M	S0304	15 mL									
Sodium chloride solution, 0.5 M	S0348				150 mL						
Sodium hydroxide solution, 0.1 M	S0149						10 mL				
Sodium nitrate	S0281		5 g								
Sodium sulfate solution, saturated	S0373						200 mL				
Sodium sulfate solution, 1 M	S0352										
Sodium sulfate solution, 0.5 M	S0353	500 mL									
Sodium thiosulfate, pentahydrate	S0114				372 g						
Starch solution, 0.5%	S0151				15 mL						
Steel wool	S0128		1								
Sucrose	S0134									50 g	
Sulfuric acid solution, 1 M	S0202				250 mL	200 mL					
Tin strips	T0087										1

Continued on next page

(for a class of 30 students working in pairs)   **Experiments and Demonstrations**

	Flinn Scientific Catalog No.	Introduction to Electrochemistry	Measuring Cell Potentials	Quantitative Electrochemistry	Electrolysis Reactions	Microscale Electrolysis	Hoffman Electrolysis	The Tin Man	Orange Juice Clock	Basic Electrophoresis	Lemon Battery Contest
Chemicals, continued											
Tin(II) chloride	S0227							23 g			
Zinc strips	Z0024		15 cm								
Zinc sulfate solution, 1 M	Z0031		15 mL								
Glassware											
Beakers											
50-mL	GP1005	15									
100-mL	GP1015			15							
250-mL	GP1020									1	
400-mL	GP1025			6			1				
600-mL	GP1030					2					
1-L	GP1040			1							
Graduated cylinder, 10-mL	GP2005						1				
Petri dishes	GP3019	15									
Stirring rod	GP5075				15					1	
Test tubes, 15 × 125 mm	GP6015						2				
U-shaped tube	AP1120	15									
General Equipment and Miscellaneous											
Alligator cords	AP6052			45			3				4
Ammeter (0-1A)	AP9045			15			1				
Balance, centigram (0.01-g precision)	OB2059			3							
Barometer	AP5070						optional				
Batteries, 9-V	AP1430	15		15	1			1			
Batteries, 6-V	AP1429						1				
Batteries, size D	AP1425			60							
Battery cap with alligator clip leads	AP8954	15		15	1			1			
Battery pack	AP5621			15							
Bunsen burner	AP5344						1				
Clamp, buret	AP1034	30									
Copper wire, insulated	AP4716					22 cm					
Dual electrophoresis apparatus	FB1714									1	
Dual power pack	FB0316									1	
Electrical tape	AP6011					1					

Continued on next page

Master Materials Guide

(for a class of 30 students working in pairs)

Experiments and Demonstrations

	Flinn Scientific Catalog No.	Introduction to Electrochemistry	Measuring Cell Potentials	Quantitative Electrochemistry	Electrolysis Reactions	Microscale Electrolysis	Hoffman Electrolysis	The Tin Man	Orange Juice Clock	Basic Electrophoresis	Lemon Battery Contest
General Equipment and Miscellaneous, continued											
Electrochemical Clock Kit	AP8718								1		
Filter paper, 9 cm	AP3102		15								
Forceps, 4 inch	AP8328		15	15							
Hoffman Apparatus	AP5439						1				
Hot plate	AP4674									1	
LabPro™ Interface System	TC1500		15								optional
Latex tubing, 1/8" I.D.	AP2076						4 cm				
Logger Pro software	TC1421	1									optional
Metal electrode set	AP4602										1
Micro pipet, digital	AP1804									1	
Micro pipet tips, small	AP1807									3	
Multimeter —or— Voltage Probe	AP4639 TC1506		15								1
Pencil leads, 0.9-mm	AP1817	30		30	1						
Petri dishes, disposable, 15 × 100 mm	AB1470							1			
Petri dishes, disposable, 3 partitions	AB1472			15							
Piezoelectric igniter	AP6609					1					
Pipets, Beral-type, graduated	AP1721	15	90	45		1					
Pipets, Beral-type, extra large bulb	AP1720					7					
Pipets, Beral-type, super jumbo	AP8850					1					
Power supply, low voltage, DC	AP9279						optional				
Scissors, heavy duty	AP8949			1							
Scissors, student	AP5394		15								
Support stand	AP8226	15									
Syringes, 10-mL	AP1730						2				
Test tube rack	AP1677					1					
Thermometer, digital	AP8716						1			1	
Timer	AP8843			15							
Wash bottle	AP1668		15	15	15						
Water, distilled or deionized	W0007 W0001	✓	✓	✓							
Wax pencils	AP8291			15							
Wooden splints	AP4455					1					

Flinn ChemTopic™ Labs — Electrochemistry